Advances in Industrial Control

Advances in Industrial Control is a series of monographs and contributed titles focusing on the applications of advanced and novel control methods within applied settings. This series has worldwide distribution to engineers, researchers and libraries.

The series promotes the exchange of information between academia and industry, to which end the books all demonstrate some theoretical aspect of an advanced or new control method and show how it can be applied either in a pilot plant or in some real industrial situation. The books are distinguished by the combination of the type of theory used and the type of application exemplified. Note that "industrial" here has a very broad interpretation; it applies not merely to the processes employed in industrial plants but to systems such as avionics and automotive brakes and drivetrain. This series complements the theoretical and more mathematical approach of Communications and Control Engineering.

Indexed by SCOPUS and Engineering Index.

Proposals for this series, composed of a proposal form (please ask the in-house editor below), a draft Contents, at least two sample chapters and an author cv (with a synopsis of the whole project, if possible) can be submitted to either of the:

Series Editor

Professor **Michael J. Grimble:**
Department of Electronic and Electrical Engineering, Royal College Building, 204 George Street, Glasgow G1 1XW, United Kingdom;
e-mail: m.j.grimble@strath.ac.uk

or the

In-house Editor

Mr. **Oliver Jackson:**
Springer London, 4 Crinan Street, London, N1 9XW, United Kingdom;
e-mail: oliver.jackson@springer.com

Proposals are peer-reviewed.

Publishing Ethics

Researchers should conduct their research from research proposal to publication in line with best practices and codes of conduct of relevant professional bodies and/or national and international regulatory bodies. For more details on individual ethics matters please see: https://www.springer.com/gp/authors-editors/journal-author/journal-author-helpdesk/publishing-ethics/14214

José M. Maestre · Carlos Ocampo-Martinez
Editors

Control Systems Benchmarks

Springer

Editors
José M. Maestre
Departamento de Ingeniería de Sistemas y Automática
Universidad de Sevilla
Sevilla, Spain

Carlos Ocampo-Martinez
Department of Automatic Control (ESAII)
Universitat Politècnica de Catalunya
Barcelona TECH
Barcelona, Spain

ISSN 1430-9491 ISSN 2193-1577 (electronic)
Advances in Industrial Control
ISBN 978-3-031-76311-3 ISBN 978-3-031-76312-0 (eBook)
https://doi.org/10.1007/978-3-031-76312-0

This Springer imprint is published by the registered company Springer Nature Switzerland AG
The registered company address is: Gewerbestrasse 11, 6330 Cham, Switzerland

To my wife Kazumi and our five daily little benchmarks:
Hiro, José, Noah, Mirai and Kenzo.

—Pepe

To Lissa and Nahia

—Carlos

Series Editor's Foreword

Control engineering is viewed differently by researchers who produce general theories and engineers who must design, calibrate, implement, and maintain control systems. Researchers usually develop theories and algorithms for control problems having a well-defined mathematical basis, and they are applicable to a range of systems represented by the type of mathematical models used. Engineers are faced with systems that are usually very specialized, and only after many assumptions and simplifications can they be designed using the general design results that are available. Engineers may also have other concerns over the limitations of equipment, quality of control, safety, security, and system downtime. The monograph series on *Advances in Industrial Control* (AIC) addresses some of the practical issues that arise and encourages the use of advanced control techniques when they can offer genuine real-world benefits.

The rapid development of new control theory, techniques, and technology has an impact on all areas of engineering. This series focuses on applications that may stimulate the development of new more general control paradigms. This is desirable if the different aspects of the "control design" problem are to be explored with the same dedication that "analysis and synthesis" problems have received. The series enables researchers to introduce new ideas motivated by challenging problems in the applications of interest. It raises awareness of the various benefits that advanced control systems design methods can provide while also explaining the challenges that can arise.

This AIC series monograph is on the topic of system benchmarks, and the text is edited by José M. Maestre and Carlos Ocampo-Martinez. The subject of benchmarking is important practically since there is a need to judge performance before systems can be improved and optimized. Various ways to judge performance are therefore needed, and this became the subject of much research about two decades ago. Given methods to judge performance, it is then useful to have application examples that can be used to assess and compare new design philosophies with classical solutions or competing advanced methods. This text includes a remarkable set of benchmark application problems with contributions from leading researchers in control.

Chapter 1 by the volume editors provides a short general introduction to the text. Chapter 2 considers a benchmark for the electricity network of the European Economic Area. This is a topic where benchmarks are needed since there are significant challenges because of the size and complexity of networks and because of the rapid shift to renewable energy sources. The benchmark treated in Chap. 3 is also concerned with the environment since it describes a full-scale demonstrator for an autonomous, self-sufficient, and sustainable greenhouse.

Chapter 4 considers the temperature control of a shrink tunnel, and it is a good example of the rather special problem applications engineers are faced with that include multiple heating zones. There may be lessons here for the metal-processing industry in the design of reheat furnace controls. Chapter 5 involves the model of a gas humidification process that is used in the process of conditioning the supply gases for a fuel-cell, a topic of much current interest.

Chapter 6 considers a benchmark for the optimization and control of wastewater treatment plants. It is one of the processes that might benefit greatly from the availability of a benchmark, since there is every indication that the water industry is not always following best practice, and it has a reputation for being slow to innovate. Chapter 7 involves the healthcare topic of automated insulin delivery in Type 1 diabetes. There is clearly a need to achieve the best control, and a benchmark has real value to determine the most suitable strategies for the wide range of patients involved. Chapter 8 involves predictive maintenance that is of great interest in industry. The particular area of a ball bearing testbench is also relevant to many transport and industrial problems.

The final chapters of the text are concerned with applications in robots and vehicles. Chapter 9, on fault-tolerant control of a quadrotor, covers a difficult control problem, where the popular nonlinear model predictive control methods have value. Such algorithms are computationally demanding, and benchmarking is a valuable tool in achieving the most effective embedded control solutions. Chapter 10 considers a multi-agent robotic system. There is a lot of activity in the theory of multi-agent systems. A benchmark for a cooperative robotic system will provide a practical example that can be used to study and understand the value of these theoretical results.

Chapter 11 involving a digital twin reflects the strong current interest in this topic. Mobile robots are also a subject of great interest, and the large number of autonomous vehicles that may be used in, say, a warehouse can make the investment in using a digital twin for benchmarking worthwhile. Chapter 12 on benchmarking for handling performance in electric vehicles using advanced control techniques is one of the topics that will receive a lot of attention. This is because of the rapid move to battery-powered vehicles, and consequent problems with vehicle handling and stability caused by the increased weight of a vehicle.

Chapter 13 is concerned with a benchmark problem for a visual nudge in the control of vehicles driven by humans. This general area is also of great current interest since in the future more advice or instructions are likely to be issued to drivers to improve safety and vehicle performance. Greater number of sensors and data-driven methods will provide the information, but the handling of this information and the way humans are alerted need careful study in which a benchmark can be exploited.

This text includes a valuable set of benchmark examples. It is valuable since this topic has probably not received the attention it really deserves. The contributions should be helpful to control-design engineers, researchers in control and instrumentation, and engineering/IT course students wishing to compare new control designs or methods. The editors and the set of leading authors have provided a very useful and welcome addition to the series.

Glasgow, UK
August 2024

Michael J. Grimble

Foreword

It is with great pleasure that I write the foreword for "Control Systems Benchmarks", edited by my colleagues José M. Maestre and Carlos Ocampo-Martinez. This comprehensive compilation is a timely contribution to the field of control engineering.

Control engineering is a discipline that deals with practical applications and real-world problem-solving. Benchmarks serve as useful tools in this domain, allowing to evaluate the performance of control systems and compare different methodologies. This book presents a collection of benchmarks that reflect the wide challenges and innovations in control engineering today. The organization into two distinct parts—Processes, and Robots and Vehicles—ensures that readers are exposed to a broad spectrum of applications and control problems.

The first part dealing with process control introduces seven benchmarks that span from load frequency control in power grids to wastewater treatment plants. Each chapter provides a detailed exploration of specific control challenges, offering insights into both the theoretical and practical aspects of process control. These benchmarks not only highlight the diversity of control problems but also present the innovative solutions being developed to address them.

The second part focuses on robotics and vehicles, presenting five benchmarks that includes fault-tolerant control of quadrotors, multi-agent strategies in robotic parks, and the lateral stability of electric vehicles, among others. These chapters illustrate the dynamic and rapidly evolving nature of control systems in robotics and vehicular applications. The inclusion of both single-robot and multi-agent systems will enrich the reader's understanding of contemporary control challenges.

Each benchmark is accompanied by either access to realistic simulators or instructions for in situ testing, enabling readers to test their strategies on the benchmarks. This hands-on approach bridges the gap between theoretical knowledge and practical application, fostering a deeper comprehension of control systems.

The contributors to this book are leading experts in their respective fields, and their collective expertise is evident in the depth and quality of the content. The chapters are written with clarity and precision, making complex concepts accessible to a wide audience, from students and researchers to professionals. The length of each chapter also ensures that the book is not only informative and easy to read.

I think that the book will inspire control students, researchers, and practitioners to apply their control techniques to new problems or use some of the methodologies used to their systems. I congratulate the editors for their remarkable effort in bringing this book to life and for their dedication to advancing the field of control engineering. It is an honor to endorse this work, which I am confident will become a valuable resource for our community.

July 2024

Prof. Eduardo Fernandez Camacho
IEEE Life Fellow, IFAC Fellow
University of Seville
Seville, Spain

Acknowledgements

This work has been partially supported by the following Spanish projects:

- The Spanish project entitled *Sustainable learning-based Management of Multi-resource Large-scale Systems* (SEAMLESS), with reference number PID2023-148840OB-I00, funded by *Ministerio de Ciencia, Innovación y Universidades* and the *Agencia Estatal de Investigación* (10.13039/501100011033/FEDER, UE).
- *Supervision and fault-tolerant control of smart infrastructures based on advanced learning and optimization* (L-BEST), with reference number PID2020-115905RB-C21, funded by *Ministerio de Ciencia, Innovación y Universidades* and the *Agencia Estatal de Investigación* (10.13039/501100011033).
- The Spanish project entitled *Control Coalicional para la Optimización de Sistemas Ciberfísicos: Ronda 3* (C3PO-R3), with reference number PID2023-152876OB-I00, funded by *Ministerio de Ciencia, Innovación y Universidades* and the *Agencia Estatal de Investigación* (10.13039/501100011033/FEDER, UE).
- *Coalitional Control for Cyber-Physical system Optimization: Round 2, Digital Doubles* (C3PO-R2D2), with grant reference number PID2020-119476RB-I00, funded by MCIN/AEI/10.13039/501100011033.
- *Efficient Management of Energy Systems including Hybrid Electrochemical Energy Storage using Digitalisation Technologies* (MASHED), with reference number TED2021-129927B-I00, funded by *Ministerio de Ciencia, Innovación y Universidades*, the *Agencia Estatal de Investigación* (10.13039/501100011033), and by the "European Union NextGenerationEU/PRTR".

Contents

Contributors

Hernán Alvarez Universidad Nacional de Colombia, Medellín, Colombia

Mateo Arcila-Osorio Automatic Control Department, Universitat Politècnica de Catalunya, Barcelona, Spain

Bart De Schutter Delft Center for Systems and Control, Delft University of Technology, CD Delft, The Netherlands

Javier Diaz-Rozo Aingura IIoT, Donostia, Spain

Raquel Dormido Department of Computer Science and Automatic Control of UNED, Madrid, Spain

Clara Escorihuela-Altaba University Clinic for Diabetes, Endocrinology, Nutritional Medicine, and Metabolism, Inselspital-University Hospital Bern, University of Bern, Bern, Switzerland

Antonio Ferramosca Department of Management, Information and Production Engineering, University of Bergamo, Dalmine, BG, Italy

Mario Francisco Department of Computer Science and Automation, Universidad de Salamanca, Salamanca, Spain

Jose Ramon D. Frejo University of Seville, Seville, Spain

Francisco García-Mañas Department of Informatics, University of Almería, Almería, Spain

Jose Garcia-Tirado University Clinic for Diabetes, Endocrinology, Nutritional Medicine, and Metabolism, Inselspital-University Hospital Bern, University of Bern, Bern, Switzerland

Juan Diego Gil Department of Informatics, University of Almería, Almería, Spain

Rubén A. González Department of Informatics, University of Almería, Almería, Spain

María Guinaldo Department of Computer Science and Automatic Control of UNED, Madrid, Spain

Ahmed Hassan University of Seville, Seville, Spain

Kenta Hoshino Graduate School of Informatics, Kyoto University, Sakyo-ku, Kyoto, Japan

Masaki Inoue Keio University, Yokohama, Kanagawa, Japan

Luca Laurenti Delft Center for Systems and Control, Delft University of Technology, CD Delft, The Netherlands

José M. Maestre University of Seville, Camino de los Descubrimientos sn., Sevilla, Spain

Francisco José Mañas-Álvarez Department of Computer Science and Automatic Control of UNED, Madrid, Spain

Montse Meneses Department of Telecommunications and Systems Engineering, Universitat Autònoma de Barcelona, Bellaterra, Spain

Manuel Muñoz Department of Informatics, University of Almería, Almería, Spain

Vihangkumar V. Naik University Clinic for Diabetes, Endocrinology, Nutritional Medicine, and Metabolism, Inselspital-University Hospital Bern, University of Bern, Bern, Switzerland

Carlos Ocampo-Martinez Automatic Control Department, Universitat Politècnica de Catalunya, Barcelona, Spain

Toshiyuki Ohtsuka Graduate School of Informatics, Kyoto University, Sakyo-ku, Kyoto, Japan

Leandro Pitturelli Department of Management, Information and Production Engineering, University of Bergamo, Dalmine, BG, Italy

Fabio Previdi Department of Management, Information and Production Engineering, University of Bergamo, Dalmine, BG, Italy

Davide Previtali Department of Management, Information and Production Engineering, University of Bergamo, Dalmine, BG, Italy

Silvana Revollar Department of Computer Science and Automation, Universidad de Salamanca, Salamanca, Spain

Alessandro Riccardi Delft Center for Systems and Control, Delft University of Technology, CD Delft, The Netherlands

Marcela Rondon Aingura IIoT, Donostia, Spain

Sara Ruiz-Moreno University of Seville, Seville, Spain

Jorge Antonio Sánchez-Molina Department of Informatics, University of Almería, Almería, Spain

Masahiko Takeda Keio University, Yokohama, Kanagawa, Japan

Pastora Vega Department of Computer Science and Automation, Universidad de Salamanca, Salamanca, Spain

Pablo Velarde Universidad Loyola Andalucía, Seville, Spain

Mateo Vernaza Universidad Internacional del Ecuador, Quito, Ecuador

Ramón Vilanova Department of Telecommunications and Systems Engineering, Universitat Autònoma de Barcelona, Bellaterra, Spain

Chapter 1
A Collection of Benchmarks for Control Engineers

José M. Maestre and Carlos Ocampo-Martinez

1.1 Introduction

In a general way, a benchmark is a standard or reference point against which things may be compared or assessed. In several fields such as computing, finance, and manufacturing, benchmarks serve as a means to measure and evaluate performance, quality, or progress. Benchmarks provide a clear, objective framework for making comparisons, identifying best practices, and driving improvements by highlighting areas that fall short of the set standards.

In the field of automatic control, a benchmark serves as a predefined set of conditions, criteria, or standard problems against which the performance of control algorithms, systems, or strategies can be evaluated. These benchmarks help in assessing the effectiveness, efficiency, robustness, and reliability of control systems under various scenarios. They provide a common platform for comparison, enabling researchers and engineers to systematically evaluate different approaches, identify strengths and weaknesses, and drive innovation by establishing performance baselines. Benchmarks in automatic control often include simulated environments, specific control tasks, or real-world applications that reflect typical challenges found in the field. By using these benchmarks, the automatic control community can ensure consistent and objective assessment, facilitating the advancement of control technologies and methodologies.

J. M. Maestre (✉)
University of Seville, Camino de los Descubrimientos sn., 41092 Sevilla, Spain
e-mail: pepemaestre@us.es

C. Ocampo-Martinez
Automatic Control Department, Universitat Politècnica de Catalunya, 08028 Barcelona, Spain
e-mail: carlos.ocampo@upc.edu

J. M. Maestre and C. Ocampo-Martinez (eds.), *Control Systems Benchmarks*, Advances in Industrial Control, https://doi.org/10.1007/978-3-031-76312-0_1

Likewise, benchmarks are not only essential tools for evaluating the performance of control systems and comparing the effectiveness of different approaches. They are also useful for gaining a deeper understanding as current or future control engineers. Students can increase their knowledge by discovering new problems where control is applied; researchers may become familiar with new application areas for their methods beyond their current scope; likewise, professionals can discover new business opportunities by transposing products and ideas across different domains.

Several benchmarks have been widely used by the scientific community to evaluate and compare control strategies along the years. A few relevant examples are here outlined:

- *Inverted Pendulum:* The inverted pendulum is a classic benchmark problem in control theory. It involves balancing a pendulum in an upright position on a moving cart. This benchmark is crucial because it represents a fundamental problem of stabilizing an inherently unstable system, which is analogous to many real-world applications such as robotics and aerospace.
- *Aircraft Autopilot Systems:* Benchmarks for aircraft autopilot systems involve simulating flight conditions to evaluate the performance of control algorithms in maintaining stable flight, handling turbulence, and executing maneuvers. These benchmarks are important as they ensure the reliability and safety of autopilot systems used in aviation.
- *Quadrotor Control:* The control of quadrotor drones is another important benchmark. It includes tasks like hovering, trajectory tracking, and handling external disturbances. This benchmark is significant due to its applications in surveillance, delivery, and search-and-rescue missions, requiring precise and robust control mechanisms.
- *Tank Level Control:* This benchmark involves maintaining the liquid level in a tank at a desired setpoint despite changes in inflow and outflow rates. It is commonly used in process control industries. Its importance lies in its relevance to chemical, pharmaceutical, and water treatment processes where precise level control is essential for safety and efficiency.
- *Two-Wheeled Self-Balancing Robot:* Similar to the inverted pendulum, this benchmark involves stabilizing and navigating a two-wheeled robot. It tests the control system's ability to maintain balance and execute movements. This benchmark is critical for applications in robotics and personal transportation devices.

However, given the transversal feature of the automatic control discipline, benchmarks based on different case studies are worth to be considered for comparison and validation purposes when designing control strategies or even estimation approaches. To this end, the areas of application covered by the book span a diverse range of industries and fields, including robotics, automotive, process control, power systems, building automation, biomedical systems, and environmental systems. By encompassing these application fields, the book is expected to show the versatility of control systems across various domains, serving as a useful resource for researchers, practitioners, and students alike. Hopefully, some of the latter will become inspired by the book and become specialists in this field.

1.2 Contents of the Book

Considering the different types of benchmarks in this book, we have merged the chapters into the following parts:

Part I: Processes

The first part of the book is composed of seven chapters, which mostly correspond to different process control problems in the industry. The benchmarks included in this part offer the opportunity to test different types of methods, as they cover a variety of dynamics (linear and nonlinear), sizes (small and large-scale), and control goals. The chapters included in this part are listed below

Chapter 2: **Load Frequency Control in the European Economic Area Electricity Network:** This first chapter of the book is provided by authors from Delft University of Technology and deals with the instability issues caused by renewable energy sources, which can be mitigated by energy storage systems. Considering the large-scale nature of the considered system, the benchmark becomes an ideal platform to test distributed control methods, which can be compared in terms of performance and computation time against a centralized Model Predictive Control (MPC) strategy provided by the authors.

Chapter 3: **Agroconnect:** The next chapter introduces a large-scale research facility at the University of Almería, which consists of an autonomous, self-sufficient, and sustainable greenhouse that includes components such as photovoltaic panels, solar thermal fields, biomass boilers, and desalination systems. Here, control challenges range from low-level regulation problems to high-level issues such as setpoint generation and resource management.

Chapter 4: **Shrink Tunnel:** The third chapter of this part is written by authors from the University of Bergamo, who present the industrial process of a shrink tunnel composed of an industrial oven fed by a conveyor belt that transports sets of grouped bottles covered by a thin plastic film. This nonlinear, underactuated, and time-delayed control challenge aims to ensure proper temperature distribution within the oven to maximize the quality of the wrapping, considering disturbances and energy costs.

Chapter 5: **Gas Humidification:** The next benchmark is provided by authors from Universitat Politècnica de Catalunya and Universidad Nacional de Colombia, focusing on a gas humidification process in a fuel-cell assembly as a benchmark for estimation and control-oriented tasks.

Chapter 6: **Wastewater Treatment Plants:** The authors from Universidad de Salamanca and Universitat Autònoma de Barcelona collaborate to present two wastewater treatment plant simulators. The first simulator comprises the water line, with five biological reactors and the secondary settler, whereas the second one is a plant-wide platform that enhances the first simulator with sludge treatment units and a primary clarifier. These are challenging control problems due to the microbiological nature of the processes involved.

Chapter 7: **Blood Glucose Control:** The next benchmark is proposed by the University of Bern and focuses on diabetes management. The chapter addresses the complex problem of blood glucose control in Type 1 Diabetes and explores the use of continuous glucose monitoring and automated insulin delivery systems, offering a platform to evaluate different control strategies aimed at improving patients' quality of life by reducing the need for constant health-related decisions.

Chapter 8: **Ball Bearing Benchmark:** The last chapter of this part is presented by an industrial partner, Aingura IIoT. The chapter provides an accelerated degradation testbench for ball bearing systems, replicating real-world operational scenarios with differences in mechanical forces, vibrations, and thermal influences.

Part II: Robots and Vehicles

The second part of the book is composed of five chapters and deals with vehicles and robots. A variety of systems and situations are considered in these benchmarks, allowing readers to experiment with both single-robot and multi-agent systems and face control challenges related to linear and nonlinear dynamics. The chapters included in this part are

Chapter 9: **Fault-Tolerant Control of a Quadrotor:** This first chapter of the second part is provided by researchers from the University of Kyoto. It deals with the need for backup control strategies in case of a fault in one of the four rotors of this type of vehicle. Under such an event, linear models are no longer valid for control design, and nonlinear dynamics need to be accounted for.

Chapter 10: **Robotic Park:** This chapter, written by researchers from Universidad Nacional de Educación a Distancia (Spain), presents an experimental platform to test multi-agent control strategies with different types of aerial and ground robots. Remarkably, Robotic Park allows working in a virtual environment, in a physical environment, or in a hybrid scheme using ROS 2.

Chapter 11: **Autonomous Holonomic Robot:** The authors from Universidad Internacional del Ecuador and Universidad Loyola Andalucía bring a scaled low-cost prototype of a robot and a digital twin to simulate the perception and localization of the robot.

Chapter 12: **The FOX Vehicle:** The authors from Universidad de Sevilla bring FOX, a real vehicle available for tests at this university. A realistic simulator of the vehicle dynamics, implemented in Simmechanics, is also offered to assess the lateral stability of electric vehicles in turning maneuvers.

Chapter 13: **Light-Guided Control of Human-Driven Vehicles:** The last chapter of the book is provided by researchers from the University of Keio and Universidad de Sevilla, and deals with the behavior of humans as drivers. It presents a benchmark to study drivers' reactions to light signals in tunnels. The goal is to use these lights to help drivers maintain their speed as they enter the tunnel, as it is common for them to reduce their speed in such situations, causing traffic jams in severely congested roads.

1.3 Making the Most of the Book

As can be seen, each benchmark description is performed by experts in the field who have shared their experience and insights on the topic. Additionally, they have written these chapters with a nominal length of 15–20 pages to make the book agile for the reader. More importantly, they provide the means to access their benchmarks. Sometimes this will be a link where the code to run a realistic simulator can be downloaded; other times, the authors provide instructions to access their labs and test the control strategies in situ. Either way, the reader will have a formidable opportunity to expand their portfolio of control applications and network of collaborators.

Enjoy!

Part I
Processes

Chapter 2
A Benchmark for the Application of Distributed Control Techniques to the Electricity Network of the European Economic Area

Alessandro Riccardi, Luca Laurenti, and Bart De Schutter

2.1 Introduction

2.1.1 The Origin of the Benchmark: Motivation and Challenges

The European Economic Area Electricity Network Benchmark (EEA-ENB) is a benchmark designed for the implementation and testing of distributed control strategies for large-scale power networks. The idea behind the benchmark is to build an abstract model of the European network of transmission systems for electricity. We represent each country of the European economic area as an independent electrical area connected to others through tie lines according to a predefined electricity network topology. The result is a real-world oriented benchmark that accounts for the presence of renewable generation and Energy Storage Systems (ESSs) in the Load Frequency Control (LFC) problem of the power network.

The development of the EEA-ENB is essential because no established control model for the European electricity transmission system consistently serves as a reference for distributed control techniques, especially with energy storage systems and renewable energy sources. Additionally, the use case for the EEA-ENB is not restricted only to the pure development of control strategies. With minimal modifications, it can also be used for other applications, such as the economic optimization of

A. Riccardi (✉) · L. Laurenti · B. De Schutter
Delft Center for Systems and Control, Delft University of Technology, Mekelweg 2, 2628 CD Delft, The Netherlands
e-mail: a.riccardi@tudelft.nl

L. Laurenti
e-mail: l.laurenti@tudelft.nl

B. De Schutter
e-mail: b.deschutter@tudelft.nl

J. M. Maestre and C. Ocampo-Martinez (eds.), *Control Systems Benchmarks*, Advances in Industrial Control, https://doi.org/10.1007/978-3-031-76312-0_2

network operation, the study of network expansion strategies, testing of security and privacy features, and simulation of emergency situations such as cascading blackouts and network restoration.

To assess the time and computation requirements for the implementation of a distributed control strategy, we implement centralized Model Predictive Control (MPC) on the network. Together with the value of the cost function of centralized MPC developed, this provides the user metrics to evaluate the advantages and disadvantages in the implementation of a specific distributed control technique. The EEA-ENB is formulated with a modular approach such that extensions can be implemented if needed, allowing for various application scenarios as mentioned before. The stability of the network is assessed through the study of LFC problem. Moreover, another application that is particularly relevant for this benchmark is the economical optimization of energy trading among network agents. The EEA-ENB can also be employed to formulate Distributed MPC (DMPC) techniques in the presence of hybrid dynamics, thanks to a modified ESS dynamics reported. Additional extensions, not included in this work, include the characterization of each electrical area according to the deregulated energy market through the modeling of generation plants, the auction system for scheduling energy production across the various generation companies, and the market of power exchanges between different electrical areas [4].

The main challenge in controlling the EEA-ENB has to be sought in its scale: 26 electrical areas are considered, each subject to distinct variations in load requests and renewable generation. When using a growing number of control agents, the computation time of a centralized control action becomes increasingly prohibitive; thus, distributed control approaches are required.

2.1.2 Load Frequency Control in Modern Power Networks

The LFC problem is a crucial challenge in power systems, and it has a particular socio-economic interest [15]. The LFC problem gained interest in the research community in the 1970s [9] after some major system events led to cascading blackouts [12]. These problems typically arise when unexpected changes in the load of a power system occur, with consequent shifts in the operating frequency of the electrical area under consideration, and the propagation of this effect to neighboring areas. In the last decades, cascading blackouts have been exacerbated by the increasing diffusion of renewable energy sources, which are posing new challenges for LFC of interconnected power grids due to their intermittent and stochastic nature, and inertialess generation [20].

Nowadays, new strategies to increase network robustness are constantly sought [4]. This is the reason why ESSs are fundamental in modern energy grids: they allow for more efficient use of energy, optimizing its usage based on the demand, and they can be used to counteract the inertialess properties of renewable energy sources. Therefore, part of the modeling section of this chapter is dedicated to ESSs, from the simplest dynamical formulation to more complex hybrid formulations.

Formally, the main control problem solved in the EEA-ENB is the regulation to zero of the frequency deviation of the network from the nominal value. This problem is solved in the presence of unexpected changes in the load, renewable generation, and ESSs. Early approaches to the solution were mainly based on PID control theory. With the progression of technology, more advanced techniques have been implemented, such as variable gain scheduling, fuzzy logic control, artificial neural networks, and optimal control [15, 20].

In this chapter, we propose MPC as a reference control technique for the benchmark, and DMPC as its natural extension. The choice of MPC is related to the fact that it provides the optimal control action according to a certain cost function defined by the user, while incorporating constraints on the evolution of the state and control. Nevertheless, the EEA-ENB is designed to be control-independent, and virtually all control techniques can be implemented on it. For a detailed list of control approaches for the LFC problem, we refer to [20].

2.2 Problem Description

2.2.1 System Description

The EEA-ENB is composed of 26 interacting electrical areas connected through tie lines and uses real-world data acquired from the European Network of Transmission System Operators for Electricity (ENTSO-E) transparency platform accessible from [1]. Each area represents an equivalent electrical machine aggregating the inertia and dispatchable capacity of generators in that specific area, a modeling technique commonly used in the context of LFC [15]. The electrical topology of the network is derived from the grid map also provided by ENTSO-E [1]. The benchmark is constituted by 26 control areas due to considerations about the availability and scale of the data about the 31 members of the EEA. The electrical topology of the resulting network is reported in Fig. 2.1, where each country is labeled with the respective ISO code. Positions of the areas in the space are selected according to their geographical centroids, and, on this basis, the lengths of the tie lines are defined using the Euclidean distance as reported in Table 2.1. In this graph representation, each node is associated with dynamics incorporating generation, storage, consumption, and interaction behaviors of the considered electrical area and of its neighborhood. In particular, an electrical area is composed of a multiplicity of autonomous subsystems working together to guarantee the satisfaction of the setpoints assigned by the area-level controller. The aggregation of those subsystems allows one to define an equivalent electrical machine for each area. Specifically, each area may comprehend the following:

- A *dispatchable generator* used to model all sources of energy that can be actively controlled to balance the load. Conventional power sources are hydroelectric turbines, nuclear power plants, and gas, oil, or coil turbines. Those sources are asso-

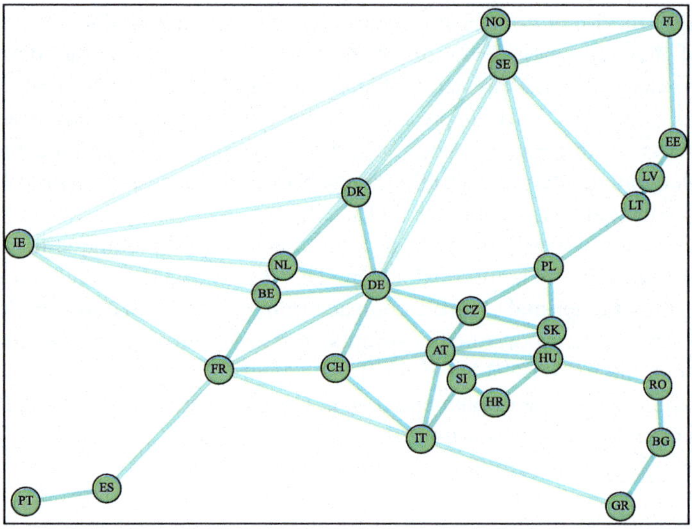

Fig. 2.1 Electrical topology of the EEA-ENB. Each node represents an electrical area, coinciding with a single country, whereas each edge is a transmission line. The transparency of the edges represents the strength of interaction among the areas

ciated with an aggregated power generation that we can allocate at each time step according to the production limits of each area.

- *Non-dispatchable generation* associated with renewable energy production, such as wind and solar generation, which have intermittent and stochastic nature. We assume that data are available both for the exact value of the produced power and for day-ahead forecasts.
- An *ESS* used to accumulate and supply energy at the best convenience and according to the control strategy implemented. In general, energy storage systems can be classified into three macro-categories: electrical storage (e.g., ultracapacitors), electrochemical storage (e.g., batteries), and mechanical storage (e.g., water reservoirs). However, this distinction is not considered in the benchmark, but it is suggested as a possible extension. Following the same approach used for dispatchable generation, we consider the aggregated storage and power of all the ESSs in the electrical area.
- A *load demand* for which measurements and day-ahead forecasts are available.

Those components contribute to the internal load frequency balance of the electrical area. Moreover, a power exchange among areas is present over the tie lines reported in the electrical topology. This interaction must also be accounted for in the overall power balance.

Table 2.1 Lengths of the tie lines expressed in 10^3 [km]

	AT	BE	BG	HR	CZ	DK	EE	FI	FR	DE	GR	HU	IE	IT	LV	LT	NL	NO	PL	PT	RO	SK	SI	ES	SE	CH
AT	0	0	0	0	2.65	0	0	0	0	4.82	0	5.61	0	4.76	0	0	0	0	0	0	0	5.88	1.85	0	0	5.58
BE	0	0	0	0	0	0	0	0	4.69	5.77	0	0	13.19	0	0	0	1.75	0	0	0	0	0	0	0	0	0
BG	0	0	0	0	0	0	0	0	0	0	4.02	0	0	0	0	0	0	0	0	0	3.01	0	0	0	0	0
HR	0	0	0	0	0	0	0	0	0	0	0	3.61	0	0	0	0	0	0	0	0	0	0	2.12	0	0	0
CZ	2.65	0	0	0	0	0	0	0	0	5.13	0	0	0	0	0	0	0	0	4.67	0	0	4.33	0	0	0	0
DK	0	0	0	0	0	0	0	0	0	5.03	0	0	17.82	0	0	0	5.44	11.56	0	0	0	0	0	0	10.22	0
EE	0	0	0	0	0	0	0	6.37	0	0	0	0	0	0	2.2	0	0	0	0	0	0	0	0	0	0	0
FI	0	0	0	0	0	0	6.37	0	0	0	0	0	0	0	0	0	0	8.99	0	0	0	0	0	0	8.89	0
FR	0	4.69	0	0	0	0	0	0	0	9.35	0	0	12.38	11.19	0	0	0	0	0	0	0	0	0	8.58	0	6.09
DE	4.82	5.77	0	0	5.13	5.03	0	0	9.35	0	0	0	0	0	0	0	4.98	15.23	9.06	0	0	0	0	0	13.41	4.84
GR	0	0	4.02	0	0	0	0	0	0	0	0	0	0	10.94	0	0	0	0	0	0	0	0	0	0	0	0
HU	5.61	0	0	3.61	0	0	0	0	0	0	0	0	0	0	0	0	0	0	0	0	5.87	1.48	4.63	0	0	0
IE	0	13.19	0	0	0	17.82	0	0	12.38	0	0	0	0	0	0	0	13.84	27.51	0	0	0	0	0	0	0	0
IT	4.76	0	0	0	0	0	0	0	11.19	0	10.94	0	0	0	0	0	0	0	0	0	0	0	3.81	0	0	5.84
LV	0	0	0	0	0	0	2.2	0	0	0	0	0	0	0	0	1.69	0	0	0	0	0	0	0	0	0	0
LT	0	0	0	0	0	0	0	0	0	0	0	0	0	0	1.69	0	0	0	5.55	0	0	0	0	0	10.14	0
NL	0	1.75	0	0	0	5.44	0	0	0	4.98	0	0	13.84	0	0	0	0	16.99	0	0	0	0	0	0	0	0
NO	0	0	0	0	0	11.56	0	8.99	0	15.23	0	0	27.51	0	0	0	16.99	0	0	0	0	0	0	0	2.28	0
PL	0	0	0	0	4.67	0	0	0	0	9.06	0	0	0	0	0	5.55	0	0	0	0	0	3.37	0	0	10.93	0
PT	0	0	0	0	0	0	0	0	0	0	0	0	0	0	0	0	0	0	0	0	0	0	0	4.34	0	0
RO	0	0	3.01	0	0	0	0	0	0	0	0	5.87	0	0	0	0	0	0	0	0	0	0	0	0	0	0
SK	5.88	0	0	0	4.33	0	0	0	0	0	0	1.48	0	0	0	0	0	0	3.37	0	0	0	0	0	0	0
SI	1.85	0	0	2.12	0	0	0	0	0	0	0	4.63	0	3.81	0	0	0	0	0	0	0	0	0	0	0	0
ES	0	0	0	0	0	0	0	0	8.58	0	0	0	0	0	0	0	0	0	0	4.34	0	0	0	0	0	0
SE	0	0	0	0	0	10.22	0	8.89	0	13.41	0	0	0	0	0	10.14	0	2.28	10.93	0	0	0	0	0	0	0
CH	5.58	0	0	0	0	0	0	0	6.09	4.84	0	0	0	5.84	0	0	0	0	0	0	0	0	0	0	0	0

2.2.2 System Dynamics

The topology of the power system is represented as a graph $\mathcal{G} = (\mathcal{V}, \mathcal{E})$ where each node $v_i \in \mathcal{V}$ is associated with an independent electrical area i, and each undirected edge $\epsilon_{ij} = \epsilon_{ji} = (v_i, v_j) \in \mathcal{E} \subseteq \mathcal{V} \times \mathcal{V}$ is a tie line connecting adjacent areas i and j, allowing for bidirectional power flow. In our case, we have 26 nodes, one for each electrical area. The presence of an edge represents the existence of a power connection. For each node v_i, we define its neighborhood as $\mathcal{N}_i = \{v_j \in \mathcal{V} \mid (v_i, v_j) \in \mathcal{E}\}$, i.e., the set of nodes connected to the node v_i. To each node v_i, $i = 1, \ldots, 26$, an equivalent electrical machine is associated according to the schematic in Fig. 2.2. Each electrical area i is always characterized by at least three states: the angle δ_i [deg] of the rotor, the operating frequency f_i [Hz] of the equivalent machine, and the energy e_i [GWh] stored in the ESS. The control inputs for the i-th area are the deviation ΔP_i^{disp} [GW] of dispatchable power production w.r.t. the scheduled value, and the power $P_i^{\text{ESS, c}}$ [GW] supplied to or $P_i^{\text{ESS, d}}$ [GW] withdrawn from the ESS. Additionally, each area is subjected to the influence of external inputs: the variation in the load request ΔP_i^{load} [GW], renewable energy production ΔP_i^{ren} [GW], and the power transmitted over the tie lines ΔP_i^{tie} [GW] connected to area i.

For this system, it is common to assume [4, 15] a linearized discrete-time model around an operating point $(\delta_{0,i}, f_{0,i})$ for the power angle and frequency dynamics for each area i. For all $i \in \{1, \ldots, 26\}$, we assume $\delta_{0,i} = 30$ [deg], but this depends on the scheduled power exchanges among electrical areas as specified later, with limits $\delta_0 \in (0, 90)$ [deg]; moreover, the operating frequency of the European power network is $f_{0,i} = 50$ [Hz] [15]. Regarding the ESS of the i-th area, the simplest model capturing the charging and discharging characteristics of a storage system is the linear representation also reported in [22]. Extensions of this model, and alternative

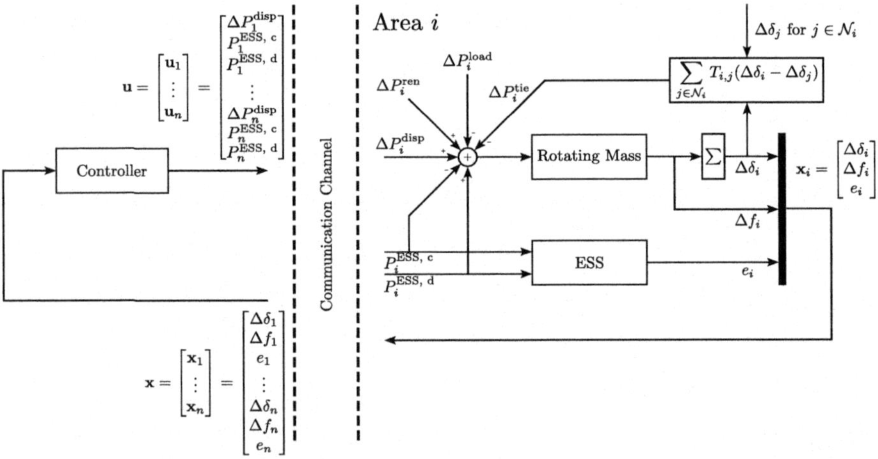

Fig. 2.2 Schematic of the network of equivalent machines, with details of the i-th electrical area

formulations of the ESS dynamics, as the PWA description in [18], are discussed in Sect. 2.2.4. For the tie line interaction, we also use a linearized equation [15] under the assumption that machine angle deviations are small enough, which will be guaranteed through operating constraints in the control formulation.

To summarize, the state, input, and external input of the i-th area are the vectors:

$$\mathbf{x}_i = \left[\Delta\delta_i \ \Delta f_i \ e_i\right]^\mathsf{T} \quad \mathbf{u}_i = \left[\Delta P_i^{\text{disp}} \ P_i^{\text{ESS, c}} \ P_i^{\text{ESS, d}}\right]^\mathsf{T} \quad \mathbf{w}_i = \left[\Delta P_i^{\text{load}} \ \Delta P_i^{\text{ren}} \ \Delta P_i^{\text{tie}}\right]^\mathsf{T}, \tag{2.1}$$

and their aggregation provides the respective definition of the state, input, and external input vectors for the overall network:

$$\mathbf{x} = \left[\mathbf{x}_1^\mathsf{T} \cdots \mathbf{x}_{26}^\mathsf{T}\right]^\mathsf{T} \quad \mathbf{u} = \left[\mathbf{u}_1^\mathsf{T} \cdots \mathbf{u}_{26}^\mathsf{T}\right]^\mathsf{T} \quad \mathbf{w} = \left[\mathbf{w}_1^\mathsf{T} \cdots \mathbf{w}_{26}^\mathsf{T}\right]^\mathsf{T}. \tag{2.2}$$

Assuming that the discrete-time dynamics obtained through forward Euler discretization has sampling time τ, the dynamics of the i-th electrical area has the form:

$$\mathcal{S}_i : \begin{cases} \Delta\delta_i(k+1) = \Delta\delta_i(k) + \tau 2\pi \Delta f_i(k) \\ \Delta f_i(k+1) = \left(1 - \frac{\tau}{T_{\text{p},i}}\right)\Delta f_i(k) + \tau\frac{K_{\text{p},i}}{T_{\text{p},i}}\big(\Delta P_i^{\text{disp}}(k) - \Delta P_i^{\text{load}}(k) + \Delta P_i^{\text{ren}}(k) \\ \qquad -\Delta P_i^{\text{tie}}(k) - P_i^{\text{ESS, c}}(k) + P_i^{\text{ESS, d}}(k)\big) \\ e_i(k+1) = e_i(k) + \tau\left(\eta_i^{\text{c}} P_i^{\text{ESS, c}}(k) - \frac{1}{\eta_i^{\text{d}}} P_i^{\text{ESS, d}}(k)\right) \end{cases} \tag{2.3}$$

$$\Delta P_i^{\text{tie}}(k) = \sum_{j \in \mathcal{N}_i} T_{ij}(\Delta\delta_i(k) - \Delta\delta_j(k)), \tag{2.4}$$

where $K_{\text{p},i}$ and $T_{\text{p},i}$ are respectively the gain and the time constants of the dynamics of the rotating mass; η_i^{c} and η_i^{d} are charging and discharging rates of the battery with $0 < \eta_i^{\text{c}}, \eta_i^{\text{d}} < 1$; and T_{ij} [GW/deg] is the gain associated with the tie line (i, j), i.e., $T_{ij} = k_{ij}/d_{ij}$, which depends on the geographical distance d_{ij} [km] among the electrical areas, and on the gain k_{ij} [km·GW/deg], which is assumed to be equal to 1 for all i, j in this chapter.

2.2.3 Assumptions and Operating Conditions

The electrical angle deviation is constrained as $-30 \le \Delta\delta_i \le 30$, so that the electrical angle satisfies $0 \le \delta_i \le 60$, with $\delta_{0,i} = 30$ [deg]. For the operating frequency, we assume the range $-0.04 \le \Delta f_i \le 0.04$, with $f_{0,i} = 50$ [Hz] [4, 15]. For the ESSs, we consider the maximum storage capacity to be equal to the total dispatchable capacity, i.e., $0 \le e_i \le e_{i,\max}^{\text{disp}}$, for each area i, with $e_{i,\max}^{\text{disp}} = 1 \cdot P_{i,\max}^{\text{disp}}$ [GWh]. For each electrical area i the following state constraints hold:

$$-30 \le \Delta\delta_i \le 30 \quad -0.04 \le \Delta f_i \le 0.04 \quad 0 \le e_i \le P_{i,\max}^{\text{disp}}. \tag{2.5}$$

Input limits are selected such that the total available dispatchable or storage capacity can be allocated over one hour:

$$-\frac{P_{i,\max}^{\text{disp}}}{1440} \leq \Delta P_i^{\text{disp}} \leq \frac{P_{i,\max}^{\text{disp}}}{1440} \qquad 0 \leq P_i^{\text{ESS, c}}, \; P_i^{\text{ESS, d}} \leq \frac{P_{i,\max}^{\text{disp}}}{1440}. \tag{2.6}$$

The sampling time of the system is $\tau = 2.5$ [s], which is 10 times faster than the time constant $T_{\text{p},i} = 25$ [s]. A variation in the external inputs occurs every 1440 time steps, i.e., every hour. A simulation of $24 \cdot 1440 = 34560$ steps would use 24 hours of real-world data about load and renewable generation; see Sect. 2.3.1 for additional details.

2.2.4 Extensions and Alternative Formulations

We propose three directions to modify or extend the proposed dynamics: a PWA formulation of the ESS dynamics, an extension to include the behavior of turbines and pumps, and an augmented state representation to describe the energy market. Other possible extensions are reported at the end of this section.

ESS Hybrid Dynamics

Assuming that the ESSs can only be in a charging or discharging state at each time step, their dynamics can be described with the following Piecewise Affine Linear (PWA) equations [18]:

$$e_i(k+1) = \begin{cases} e_i(k) + \tau \eta_i^{\text{c}} \Delta P_i^{\text{ESS}}(k) \text{ if } \Delta P_i^{\text{ESS}}(k) \geq 0 \\ e_i(k) + \tau \frac{1}{\eta_i^{\text{d}}} \Delta P_i^{\text{ESS}}(k) \text{ if } \Delta P_i^{\text{ESS}}(k) < 0. \end{cases} \tag{2.7}$$

In this formulation, the charging and discharging inputs used in (2.3) are substituted by a single input P_i^{ESS}, representing the total power exchange of the electrical area with the ESS. This formulation is completely different from the linear one in (2.3), and it can be demonstrated that the two representations are equivalent only if the ESS is lossless, i.e., if $\eta^{\text{c}} = \eta^{\text{d}} = 1$.

Turbine and Pump Dynamics Extension

A finer representation of the system would include the presence of a turbine for the generation of the dispatchable power, and of a turbine/pump system for mechanical ESS to allocate and use energy in the water reservoirs (this is not necessary for other types of ESSs). This concept can be applied both to the ESS formulation in (2.3) and (2.7). Additional states are introduced in this new description: the signals previously considered in (2.3) and (2.7) as inputs are now the states of the turbines or pump. Additionally, new inputs u_i^{disp}, $u_i^{\text{ESS, c}}$, $u_i^{\text{ESS, d}}$ are introduced to control the turbines and pump. Specifically, if we consider the linear formulation (2.3) for the i-th electrical area we have

$$\mathbf{x}_i = \left[\Delta\delta_i \ \Delta f_i \ e_i \ \Delta P_i^{\text{disp}} \ P_i^{\text{ESS, c}} \ P_i^{\text{ESS, d}}\right]^\mathsf{T} \quad \mathbf{u}_i = \left[u_i^{\text{disp}} \ u_i^{\text{ESS, c}} \ u_i^{\text{ESS, d}}\right]^\mathsf{T} \tag{2.8}$$

and the dynamics (2.3) is augmented with the update equations:

$$\begin{aligned} \Delta P_i^{\text{disp}}(k+1) &= \left(1 - \tfrac{\tau}{T_{\text{t},i}}\right) \Delta P_i^{\text{disp}}(k) + \tau \tfrac{K_{\text{t},i}}{T_{\text{t},i}} u_i^{\text{disp}} \\ P_i^{\text{ESS, c}}(k+1) &= \left(1 - \tfrac{\tau}{T_{\text{c},i}}\right) \Delta P_i^{\text{ESS, c}}(k) + \tau \tfrac{K_{\text{c},i}}{T_{\text{c},i}} u_i^{\text{ESS, c}} \\ P_i^{\text{ESS, d}}(k+1) &= \left(1 - \tfrac{\tau}{T_{\text{d},i}}\right) \Delta P_i^{\text{ESS, d}}(k) + \tau \tfrac{K_{\text{d},i}}{T_{\text{d},i}} u_i^{\text{ESS, d}} \end{aligned} \tag{2.9}$$

where $T_{\text{t},i}$, $T_{\text{c},i}$, $T_{\text{d},i}$ and $K_{\text{t},i}$, $K_{\text{c},i}$, $K_{\text{d},i}$ are respectively the time constants and gains of the turbine and storage turbine/pump of the i-th electrical area. As good engineering practice, the time constants $T_{\text{t},i}$, $T_{\text{c},i}$, $T_{\text{d},i}$ are selected to be at least 10 times smaller than $T_{\text{p},i}$, and accordingly the sampling time τ has to be at least 100 times smaller than the original one in (2.3), i.e., $\tau = 0.025$ [s]. For further details see [4, 7].

State Augmentation and Total Production Constraints

An aspect usually not considered in LFC systems are the constraints on the total dispatchable production. In this benchmark, we want to constrain the total dispatchable production P_i^{disp} at time step k to be non-negative, and smaller than the overall capacity of a certain area. Moreover, we might require that P_i^{tie}, which is the total energy transmitted over the tie lines connected to area i, is between certain limits, or tracks a reference. To this end, the dynamics (2.3) are augmented with the two states P_i^{disp}, P_i^{tie} that evolve as follows:

$$\begin{aligned} P_i^{\text{disp}}(k+1) &= P_i^{\text{disp}}(k) + \tau \Delta P_i^{\text{disp}}(k) \\ P_i^{\text{tie}}(k+1) &= P_i^{\text{tie}}(k) + \tau \Delta P_i^{\text{tie}}(k). \end{aligned} \tag{2.10}$$

In this way, we impose limits on the overall dispatchable generation for each electrical area by adding the constraint $0 \leq P_i^{\text{disp}}(k) \leq P_{i,\max}^{\text{disp}}$ to the control problem. For example, some areas may have a renewable production that exceeds the total load request. This constraint ensures that the excess is stored for later use or transmitted to neighboring areas. As initial condition, we assume to have a dispatchable production that compensates for the total load request and that accounts for the renewable generation:

$$P_i^{\text{disp}}(0) = \max\left\{0;\ P_i^{\text{load}}(0) - P_i^{\text{ren}}(0)\right\}. \tag{2.11}$$

Additional Extensions

The equivalent machine modeling approach can also be applied to the deregulated energy market [4]. This approach involves defining various actors for electricity production in different regions, each with its dispatchable generation capacity. These actors, known as generation companies, can be represented by individual turbines that aggregate the inertia of all the generators within the same company. Additionally, there are ESSs that aggregate the storage capacities of each company. Thus, in each area i, there is a certain number of dispatchable generators and ESSs. A centralized auction system determines which generation company supplies energy to each area, considering cross-border production, electrical topology, and predefined operational strategies.

We also highlight the fact that each electrical area can be further subdivided into frontier sectors and a central sector, with tie lines connecting them. This subdivision of the electrical topology can be used for energy trade modeling. The central sector may account for the generation of critical infrastructures and is connected to all the frontier sectors. Each frontier sector is connected to the frontier sector of a neighboring area and to the central sector of the area it belongs to. This further subdivision of the topology can be used to mitigate the effect of power transmission from adjacent areas, to ensure enhanced stability of the central sector, and to define scheduled power transmissions among neighboring areas.

Future research should also consider the exploration of different ESS technologies, the challenges related to their implementation, their feasibility, and economic sustainability, all aspects that can contribute to the further refinement of the EEA-ENB.

2.2.5 Goal of the Control System

The main control goals for the benchmark that are satisfied by the centralized MPC strategy (2.12) are the following:

- *Regulation* of the frequency deviation $\Delta f_i = f_i - f_{i,0}$ of each electrical area to zero, so that the frequency of the network stays at the desired value $f_0 = 50$ [Hz]. This is also the main goal of the control system. Moreover, we also regulate the machine angle deviation $\Delta\delta_i = \delta_i - \delta_{i,0}$ to zero, such that the efficiency of the machine is preserved.
- *Operational Constraint Satisfaction:* In addition to the regulation goals, the control system should also ensure that the operational constraints of Sects. 2.2.3 and 2.2.4 are satisfied. Ensuring that these constraints are satisfied is of primary importance for the stability of the network, and its correct functioning. Similar constraints should be enforced also for the augmented models in Sect. 2.2.4.
- *Disturbance Rejection:* From a control perspective, the external signals of load and renewable generation variations in (2.3) can be interpreted as disturbances to reject.
- *Minimization of the Control Effort:* The control inputs of the benchmark are the variation in dispatchable generation and power exchange with the ESS for each area i, namely $\mathbf{u}_i = [\Delta P_i^{\text{disp}} \; P_i^{\text{ESS, c}} \; P_i^{\text{ESS, d}}]^\mathsf{T}$. Minimizing the control effort $\mathbf{u}_i^\mathsf{T}\mathbf{u}_i$ for each area i while ensuring the correct functioning of the system is another relevant feature for control design.

In addition to the main control goals, other control objectives can be designed depending on the study that one wants to conduct on the benchmark. For example, if an economic MPC problem is formulated using electricity prices, then the total monetary cost for running the network can be considered, defined in terms of €·MW for each agent, for each energy source, and at each time step. With this approach, the least expensive network operation strategy can be defined, trading off the lower operational cost of the network with its stability. In this regard, if soft constraints are implemented to limit the frequency deviation, then the total-time-spent and the average-time-spent outside the optimal operation interval of the frequency can also be considered as a performance indicator. Moreover, other control goals can be considered that are more specific to the technological implementation of the network. For example, those could regard the number of charging and discharging cycles of the batteries, or their average charge level. This means that we might incorporate operational and maintenance costs in the benchmark and consider them as a way to compare control strategies.

2.3 Benchmark Design

2.3.1 Input Data

The network of equivalent machines is modeled using real data about load requests, renewable generation, and dispatchable capacities of the 26 European states selected for the implementation. Data is acquired from the ENTSO-E electricity transparency platform [1]. As an example, data for the 24 h of January 1, 2022, are reported in Fig. 2.3.

Raw data is available every hour for each area considered, both for measurements and day-ahead forecasts.

2.3.2 Implementation Details

Table 2.2 reports the parameters used in the benchmark. Their selection is done according to the parameters used in similar simulation designs [4, 7, 15]. The sampling time of the systems is selected as $\tau = 2.5$ [s]. It follows that, for each hour, i.e., for each new data sample, 1440 steps of duration τ are considered in the control simulation. We use linear interpolation to compute the external inputs $\Delta P_i^{\text{load, meas}}(k)$, for $k = 0, \dots, 34559$, from the data of $P_i^{\text{load, meas}}(h)$ available every hour, for $h = 1, \dots, 24$. The same approach is used for renewable generation measurements $P_i^{\text{ren, meas}}$, and for the forecasts $P_i^{\text{ren, for}}$, $P_i^{\text{ren, for}}$. The resulting signal variations are in Fig. 2.4.

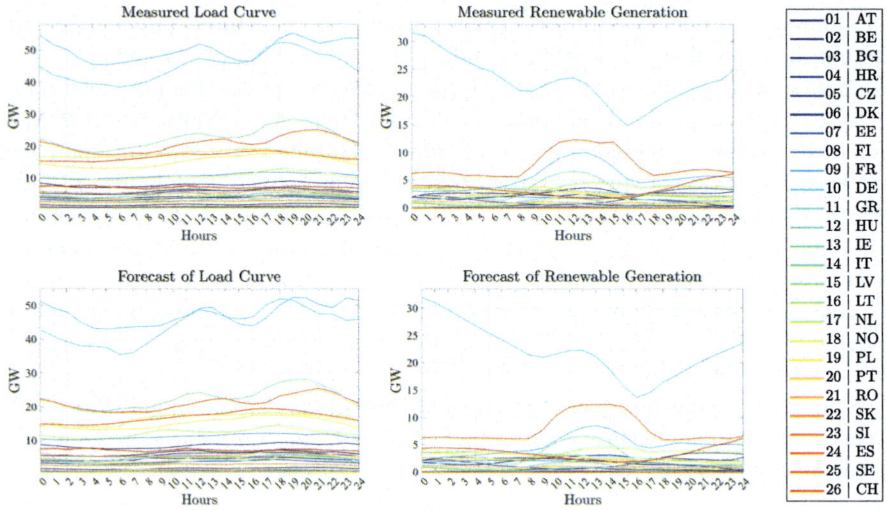

Fig. 2.3 Measurements and forecasts of the load request and renewable generation of each electrical area

Table 2.2 Parameters in the EEA-ENB

τ	$T_{\mathrm{p},i}$	$K_{\mathrm{p},i}$	η_i^{c}	η_i^{d}
2.5 [s]	25 [s]	0.05 $\left[\frac{\mathrm{Hz}}{\mathrm{GW}}\right]$	0.9	1.1

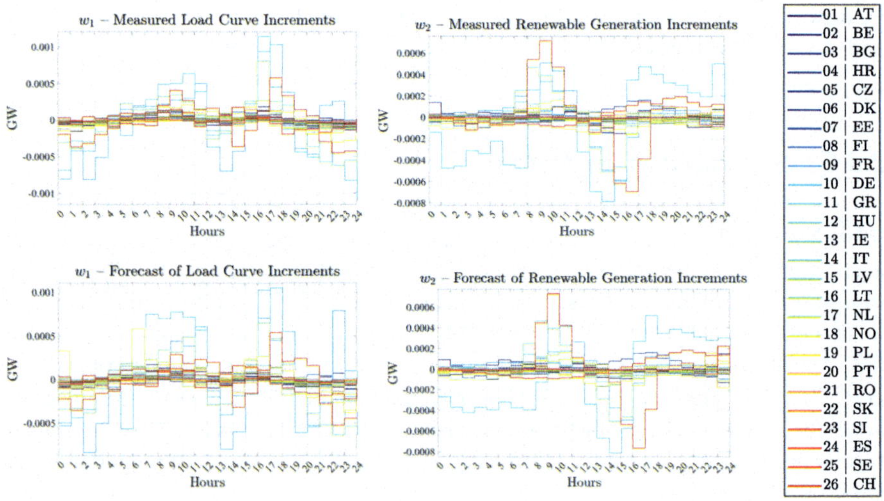

Fig. 2.4 Measurements and forecasts for the external signals

2.3.3 *Comparison with Other Benchmarks in the Field*

Several benchmarks for the simulation of power networks are present in the literature. Among the most popular ones, we can report the various implementations of IEEE buses [2]. Those benchmarks are oriented toward the simulation of power networks for electrical engineering applications. The benchmark we propose is instead oriented to the implementation of distributed control techniques, with predictive control as primary objective. To the best of our knowledge, an LFC-oriented benchmark modeled using data from the EEA is not present in the literature. A similar study was performed in [7] for the simulation of the Northern European network, but without considering renewable generation and ESSs.

2.3.4 *Performance Metrics*

The performance of the control system is measured using the value of the quadratic cost function that will be formally introduced in the control problem (2.12) of Sect. 2.5.1. Since the EEA-ENB is structured for the implementation of distributed control techniques, the overall cost over one day of simulation for the network controlled with the centralized MPC architecture (2.12) represents the optimality target for every alternative control formulation.

Another performance indicator of the control strategy is the computation time required to obtain the control action. Specifically, centralized MPC might not be suited for real-time control of the EEA-ENB due to the excessive computation time required to obtain the optimal control action. Distributed architectures are usually faster in obtaining the control law since they distribute the computational burden among the control agents, but they are more complex to implement.

As a part of the benchmark, we provide the data and results for a control simulation of the network using the centralized MPC scheme (2.12) for one day of operation of the network. The stage cost at each step, the overall cost for one day, and the total computation time are provided as a reference for alternative control strategies. Further details are in Sect. 2.5. The end-user can consider these indices as provided for a direct comparison, or perform a centralized MPC simulation with a different cost function to use personalized metrics.

2.3.5 *Alternative Test Cases*

We propose the simulation for a single day of operation of the network. However, data is available at [1] for every day of the year. Seasonality plays an important role in power generation from renewable sources. For example, solar production can increase or decrease depending on the presence of clouds, the temperature, and the length of the day. Load data is also affected by seasons. Evaluating the network with the average data about load and renewable generation of the four different seasons will give a clear view of the effectiveness of the control strategy considered over an entire year, with a mitigated variability introduced by a single day selection.

2.3.6 Output Data

Executing the benchmark will provide data about the electrical machine angle deviations, their frequency deviations, the energy stored in each area, the total power production and exchange with the ESSs, and the power transmitted over the tie lines. Those quantities are used to compute the performance metrics, and to evaluate the control strategy. Thus, both the evolution of the states of the system and the control actions can be collected and stored.

2.3.7 Essential Properties

The constraints provided for the frequency operation are essential for the stability of the network. Any prolonged deviation from the intervals provided will lead to emergency operation modes or failure of components, which in turn may generate cascading blackouts in the network. The implementation of soft constraints on the state can allow for this deviation outside safety margins, but always consider the stability of the operation of the network and the economic cost of such deviations. For more information, we refer the reader to [7, 12].

Regarding the MPC implementation, both the feasibility and stability properties should be met [3, 17]. Moreover, for the robustness of the system to the disturbances, which are the variations of the load and the renewable generation, an in-depth analysis of their evolution over an extended time window should be performed to characterize them correctly. Then, robust MPC synthesis methods could be used to guarantee this property [3].

2.4 Accessing the Benchmark

2.4.1 Links to Sources, Limitations, Costs, and Licensing

The benchmark is implemented in MATLAB® (r2023b), and the necessary files to execute it are available at [21]. Gurobi Optimizer[1] is required for the computation of

[1] https://www.gurobi.com.

the centralized MPC strategy. Alternatively, the MATLAB® Optimization Toolbox[2] can be used with minimal modifications.

The benchmark is provided for free as it is under the MIT license. Data from [1] used for the construction of the benchmark is publicly available under the Creative Commons Attribution 4.0 International License (CC BY 4.0).

2.4.2 *Documentation*

Documentation for the benchmark is available at [21]. In the following, we provide the user with information about the functions used. The data from [1] has been preliminarily checked for integrity, replacing missing entries with linear interpolations, and reported on a consistent scale. This process was performed with specialized scripts reported in the online documentation. The resulting preprocessed dataset is also part of the benchmark and provided as .csv files. The benchmark is constituted by the following files:

- `main.m`: this is the principal script to run the control simulation.
- `data_import.m`: this script reads the preprocessed data about load demands and renewable generation measurements and forecasts stored in .csv files, and returns the parameters and signals required for the simulation.
- `state_update_network.m` and `state_update_model.m`: they are identical files in the first formulation, but might be distinguished later to implement model mismatches or parameters inaccuracies. The former is used to simulate the system dynamics, and the latter as a prediction model for the centralized MPC strategy.
- `objective_function.m`: this function takes as inputs the parameters, the current value of the state, and the inputs and external inputs over the prediction window to return the total cost over that window.
- `plot_results.m`: it is used to produce plots of the simulation results and input data.

2.5 Discussion for Future Comparison

2.5.1 *Reference Approach: Centralized Predictive Control*

The LFC problem has been extensively studied in the literature. As a source of references to existing approaches for its solution, we refer to the survey [20]. To provide a comparison case for the implementation of distributed control techniques, we have implemented a centralized MPC scheme [7, 18]. Considering the system

[2] https://www.mathworks.com/products/optimization.

described in (2.3), we define at each time step k the following centralized control problem:

$$\begin{aligned} \min_{\bar{\mathbf{u}}} \sum_{j=1}^{N} & \mathbf{x}^\mathsf{T}(j|k)\mathbf{R}\mathbf{x}(j|k) + \mathbf{u}^\mathsf{T}(j-1|k)\mathbf{Q}\mathbf{u}(j-1|k) \\ \text{s.t. } & \text{dynamics (2.3)} \qquad \forall v_i \in \mathcal{V} \\ & \mathbf{x}(0|k) = \mathbf{x}(k) \\ & \text{constraints (2.5)} \qquad \forall v_i \in \mathcal{V} \\ & \text{constraints (2.6)} \qquad \forall v_i \in \mathcal{V} \end{aligned} \tag{2.12}$$

where $\bar{\mathbf{u}} = [\mathbf{u}^\mathsf{T}(0|k) \cdots \mathbf{u}^\mathsf{T}(N-1|k)]^\mathsf{T}$, $\mathbf{R}$ and $\mathbf{Q}$ are diagonal cost matrices of appropriate dimensions such that, for each area i, $\mathbf{R}_i = \text{diag}[100;\ 10;\ 1]$ and $\mathbf{Q}_i = \text{diag}[1;\ 1;\ 1]$, and $N = 30$ is the prediction horizon. For the external signals $\mathbf{w}$ we use the current measurement for $j = 1$, and their forecast for the remaining time steps. According to the receding horizon logic of MPC, we apply $\mathbf{u}(0|k)$ to the system, discard the remaining control actions, and iterate. Problem (2.12) is a quadratic optimization problem; hence efficient algorithms for solving it exist in the literature. For example, the problem can be solved using an active-set or an interior point algorithm.

For the benchmark, the optimization is performed with Gurobi Optimizer using the barrier algorithm. The simulation required 206[h], 15[m], and 24[s][3] to be completed using a processor Intel Xeon E5-2637v3, with a base clock of 3.5 GHz, and coupled with 128 GB of RAM. The solution of the optimal control problem is the vector of inputs $\mathbf{u}(k)$, for $k = 0, \ldots, 34559$ (corresponding to 24 h of simulation), reported in Fig. 2.5. In the same figure, the power transmitted over the tie lines connecting electrical areas is reported too, which allows quantifying the power trade necessary across electrical areas for the optimal operation of the network from a centralized and a cooperative perspective. Interactions among electrical areas, or agents in a generalized setting, is indeed one of the critical aspects of a distributed control strategy [8, 16], and often one of the aspects characterizing the control strategy itself or the definition of the sub-networks in cooperation/competition.

The evolution of the states resulting from the sequence of inputs obtained through MPC is reported in Fig. 2.6. For the overall power balance of each country, please consult the online repository [21].

The evolution of the cost function is presented in Fig. 2.7. Its value is the metric to evaluate distributed control techniques w.r.t. the proposed centralized approach. In the application of distributed control, the performance of the system usually decreases to achieve auxiliary objectives such as improvement in the computation time required to obtain the control action, reduction in the volume of information exchanged across the network (with advantages in the sector of security too), or reliability and redundancy of control in the presence of faults or unforeseen events.

[3] The amount of time required to run the simulation is only indicative, and serves to understand the complexity of the system. The end-user should re-run the benchmark on their own system to obtain results that are comparable with the control architectures at study.

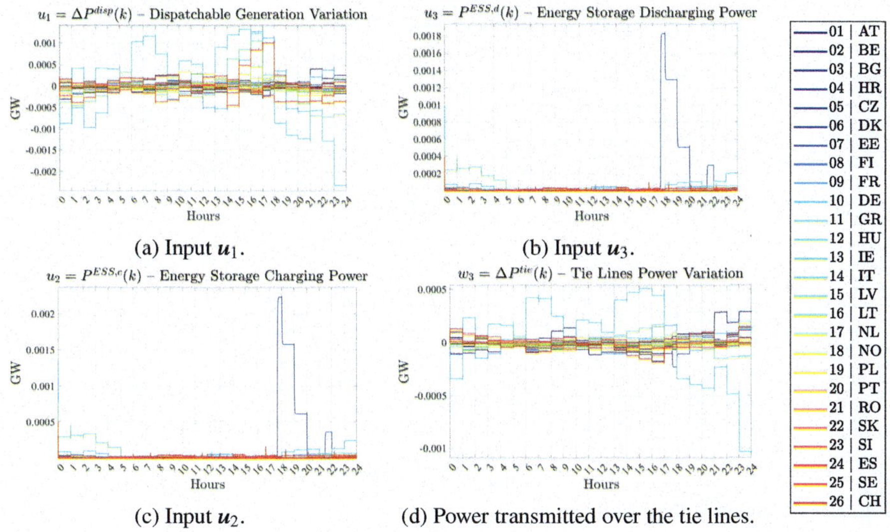

(a) Input $\boldsymbol{u}_1$.

(b) Input $\boldsymbol{u}_3$.

(c) Input $\boldsymbol{u}_2$.

(d) Power transmitted over the tie lines.

Fig. 2.5 Evolution of the inputs and of the power transmitted over the tie lines

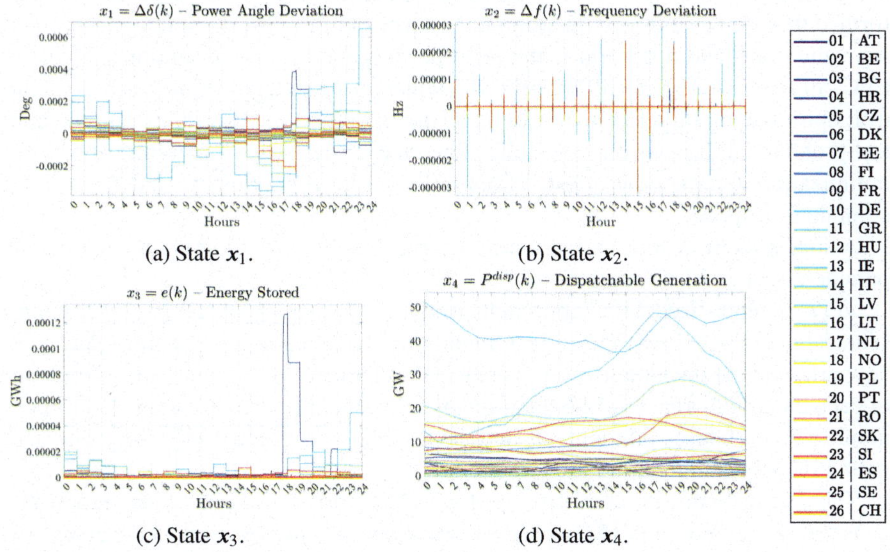

(a) State $\boldsymbol{x}_1$.

(b) State $\boldsymbol{x}_2$.

(c) State $\boldsymbol{x}_3$.

(d) State $\boldsymbol{x}_4$.

Fig. 2.6 Evolution of the states

2.5.2 Other Possible Control Approaches

To develop a deeper insight into the topics of LFC, MPC, and distributed control, some references are reported in the following. In the LFC literature, see also [20],

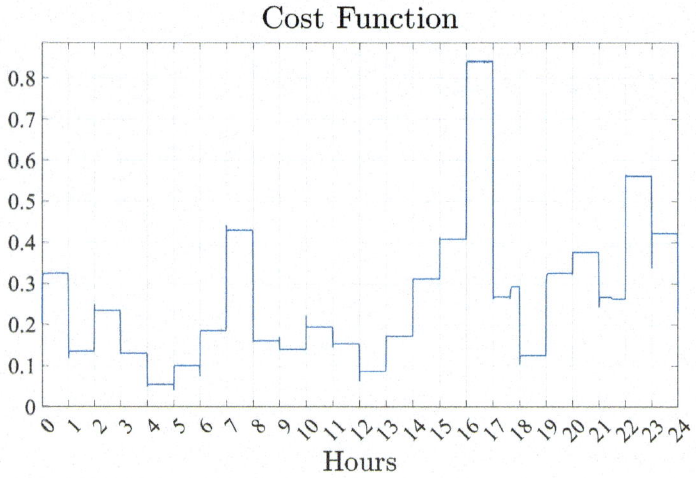

Fig. 2.7 Evolution of the cost function

many articles focus on PI control strategies where tuning of the parameters is performed through various optimization techniques [11, 14, 19]. Despite their promising validations and a general increase in performance w.r.t. conventional PI-based LFC, all these strategies still lack the fundamental advantages of model-based control: optimization of performance indices, incorporation of constraints into the control problem, ability to compensate for known external signals, and multi-objective optimization. On the other hand, PI control is easier to implement and has a faster computation speed.

A similar optimization-based tuning approach is used in [6], this time to tune the parameters of an MPC controller.

In [7] a centralized MPC approach similar to the one proposed in (2.12) is presented, but in [7], data of the Nordic transmission system is used to build an electricity network model. Power plant models are used to characterize the generation dynamics of each electrical area, but ESSs and renewable generation are not considered. Economic MPC is addressed too in [7] and the results of centralized MPC are compared with a PD controller.

The use of hydro-pumped storage for LFC of microgrids including renewable generation is explored in [5], where a control architecture based on a decentralized PD controller tuned using a Quasi-Newton optimization method is implemented.

A characterization of an electricity network using hybrid dynamics is provided in [18]. There, PWA dynamics and binary decision variables are used, leading to a hybrid MPC problem formulated as a Mixed Integer Linear Program (MILP), and solved using a branch-and-bound optimization algorithm. This approach is used to model three different aspects: the ESS, the operational mode of the microgrid, and the operational mode of the generation plants.

In [23], a distributed Linear Quadratic Regulator (LQR) is implemented to tackle the LFC problem. Methods for the distributed computation of the LQR control action are used in [23] to increase the modularity and the scalability of the control architecture. The advantages of this approach rely on the ease of implementation, and on the low computation time needed to compute the control action. The disadvantages are that the applicability of the approach is limited to linear unconstrained control problems.

A way to address Economic MPC can be found in [13]. There, both the LFC and economic load dispatch problems for power networks are considered. Those problems are usually approached using hierarchical control structures, where the economic load dispatch is at the upper level, and LFC at a lower level. Instead, in [13] the two problems are considered simultaneously, to improve the economic performance of the systems.

Another Economic MPC strategy is reported in [10]. There, a multi-objective genetic algorithm predictive control technique is used to simultaneously optimize the conflicting objectives of LFC and security-constrained economic dispatch.

2.5.3 Summary

The LFC problem has been widely investigated in the literature, and we have proposed a benchmark to evaluate distributed control strategies for solving it. The challenges arising in recent years are often related to the ever-growing use of distributed energy sources which are inertialess and can affect the frequency of the network. The use of ESSs can mitigate this effect, and we have proposed strategies to model their dynamics, extensions, and future directions for their exploration. Accordingly, future control strategies of electricity networks should account for the presence of distributed energy sources, and the implementation of distributed control strategies for efficient and resilient operation of the network. Those aspects are indeed part of the centralized MPC formulation that we have proposed as a benchmark scheme. An efficient distributed control strategy is expected to perform worse than the centralized one proposed here if only the value of the cost function is considered, but has additional properties such as a reduced computation complexity and computation time, a lower shared volume of information, or enhanced privacy, security and resilience properties.

Acknowledgements This project has received funding from the European Research Council (ERC) under the European Union's Horizon 2020 research and innovation program (Grant agreement No. 101018826)—Project CLariNet. We thank Mert Karaçelebi, Delft University of Technology, for his help in checking the modeling assumptions.

References

1. ENTSO-E: European network of transmission system operators for electricity. https://transparency.entsoe.eu/
2. IEEE DataPort. https://ieee-dataport.org
3. Bemporad A, Morari M (1999) Robust model predictive control: a survey. In: Garulli A, Tesi A (eds) Robustness in identification and control, pp 207–226
4. Bevrani H (2014) Robust power system frequency control. Springer, Berlin
5. Coban HH, Rehman A, Mousa M (2022) Load frequency control of microgrid system by battery and pumped-hydro energy storage. Water 14:1–22
6. Elsisi M, Soliman M, Aboelela MAS, Mansour W (2016) Bat inspired algorithm based optimal design of model predictive load frequency control. Int J Electri Power Energy Syst 83:426–433
7. Ersdal AM, Imsland L, Uhlen K (2016) Model predictive load-frequency control. IEEE Trans Power Syst 31:777–785
8. Fele F, Maestre JM, Camacho EF (2017) Coalitional control: cooperative game theory and control. IEEE Control Syst 37:53–69
9. Fosha CE, Elgerd OI (1970) The megawatt-frequency control problem: a new approach via optimal control theory. IEEE Trans Power Apparatus Syst 89:563–577
10. Golpîra H, Bevrani H (2014) A framework for economic load frequency control design using modified multi-objective genetic algorithm. Electric Power Compon Syst 42:788–797
11. Gupta DK, Jha AV, Appasani B, Srinivasulu A, Bizon N, Thounthong P (2021) Load frequency control using hybrid intelligent optimization technique for multi-source power systems. Energies 14:1–16
12. Haes Alhelou H, Hamedani-Golshan ME, Njenda TC, Siano P (2019) A survey on power system blackout and cascading events: research motivations and challenges. Energies 12:1–28
13. Jia Y, Dong ZY, Sun C, Meng K (2019) Cooperation-based distributed economic MPC for economic load dispatch and load frequency control of interconnected power systems. IEEE Trans Power Syst 34:3964–3966
14. Kumar R, Sharma VK (2020) Whale optimization controller for load frequency control of a two-area multi-source deregulated power system. Int J Fuzzy Syst 22:122–137
15. Kundur PS, Malik OP (2022) Power system stability and control, 2nd edn. McGraw Hill
16. Maestre JM, Negenborn RR (eds) (2014) Distributed model predictive control made easy. Springer, Berlin
17. Mayne DQ, Rawlings JB, Rao CV, Scokaert POM (2000) Constrained model predictive control: stability and optimality. Automatica 36:789–814
18. Parisio A, Rikos E, Glielmo L (2014) A model predictive control approach to microgrid operation optimization. IEEE Trans Control Syst Technol 22:1813–1827
19. Raj TD, Kumar C, Kotsampopoulos P, Fayek HH (2023) Load frequency control in two-area multi-source power system using bald eagle-sparrow search optimization tuned PID controller. Energies 16:1–25
20. Ranjan M, Shankar R (2022) A literature survey on load frequency control considering renewable energy integration in power system: Recent trends and future prospects. J Energy Storage 45:1–33
21. Riccardi A, Laurenti L, De Schutter B (2024) Code underlying the publication: a benchmark for the application of distributed control techniques to the electricity network of the european economic area
22. Sioshansi R, Denholm P, Arteaga J, Awara S, Bhattacharjee S, Botterud A, Cole W, Cortés A et al (2022) Energy-storage modeling: state-of-the-art and future research directions. IEEE Trans Power Syst 37:860–875
23. Vlahakis E, Dritsas L, Halikias G (2019) Distributed LQR design for a class of large-scale multi-area power systems. Energies 12:1–28

Chapter 3
Agroconnect Research Station: General Description and Control Challenges

Jorge Antonio Sánchez-Molina, Juan Diego Gil, Rubén A. González, Francisco García-Mañas, and Manuel Muñoz

3.1 Introduction

The European Union and the different member states are trying to promote the use of technologies that improve sustainability while increasing profitability. Despite these efforts, agriculture in general and intensive protected agriculture in particular are still not very technified. These challenges are dependent not only on aspects of crop production, but also on other issues related to fruit quality, the surrounding environment, and efficient use of resources. In this context, Agroconnect (www.agroconnect.es) aims to optimize (from an automatic control perspective) the management of resources in a self-sufficient greenhouse using renewable energies, addressing the need to link variable generation with changeable demand by applying the cyber-physical paradigm. Agroconnect is a full-scale research demonstrator infrastructure, unique in the world, consisting of a fully equipped greenhouse, which includes the generation, storage, and use of energy (electricity, heat, and cold), water, and CO_2, as shown in Fig. 3.1. This facility is located at the IFAPA (from the Spanish *Instituto Andaluz de Investigación y Formación Agraria, Pesquera, Alimentaria y de la Producción Ecológica*) station next to the University of Almería (Almería, Spain).

J. A. Sánchez-Molina · J. D. Gil (✉) · R. A. González · F. García-Mañas · M. Muñoz
Department of Informatics, University of Almería, CIESOL, ceiA3, 04120 Almería, Spain
e-mail: juandiego.gil@ual.es

J. A. Sánchez-Molina
e-mail: jorgesanchez@ual.es

R. A. González
e-mail: rgm399@ual.es

F. García-Mañas
e-mail: francisco.gm@ual.es

M. Muñoz
e-mail: mmr411@ual.es

J. M. Maestre and C. Ocampo-Martinez (eds.), *Control Systems Benchmarks*, Advances in Industrial Control, https://doi.org/10.1007/978-3-031-76312-0_3

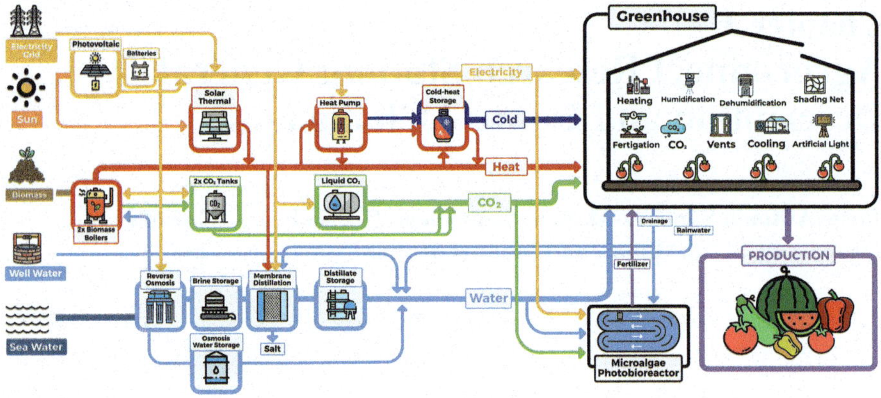

Fig. 3.1 General scheme of Agroconnect (www.agroconnect.es)

The distinctive configuration of Agroconnect, along with its array of equipment, positions it as an optimal framework for the execution and validation of diverse projects. These projects encompass a wide spectrum, ranging from digitalization, robotics, and research demonstration. Specifically, Agroconnect serves as a research demonstrator within two different living labs: (i) for desalination, in Climate Farm Demo (European Project, grant agreement 101060212), and (ii) for integration and interoperability, in the CODECs project (grant agreement 101060179). Additionally, energy, water, CO_2, and food nexus is used in other two projects, the first aimed at internalizing sustainability factors in the price of foods (FoodCost project, European Project, grant agreement 101060481), and the second devoted to developing control strategies (CyberGreen Project, National project, PID2021-122560OB-I00). Moreover, it is also used to join digital control technologies with robotics and resource-efficient practices, and it promises several key advantages in LIFE Acclimate (European Project, grant agreement 101157315). Finally, Agroconnect is also employed as a teaching tool combining virtual and augmented Reality in Agritech EU (European Project, grant agreement 101123258), and for practice in NEGHTRA (Eramus + KA2, 621723-EPP-1-2020-1-EL-EPPKA2-KA).

In particular, from a control perspective, the Agroconnect environment presents an interesting and highly intriguing framework. This is primarily attributed to the integration of numerous components and processes within the facility, each characterized by distinct dynamics and time scales. Various methodologies have been employed to tackle the diverse control challenges within the installation. These methods encompass hierarchical control, resource scheduling using the energy hubs (EH) paradigm, model predictive control (MPC), proportional–integral–derivative (PID) control, and even model-free control, among others. Such a wide range of control techniques establishes Agroconnect as a significant benchmark in the field of control systems. Moreover, Agroconnect provides the opportunity to test the practical aspects of control solutions within an actual agro-industrial facility, a crucial step for the proper validation of the controllers. This, in turn, positions Agroconnect as a vital benchmark for evaluating control strategies in the realm of large-scale control applications and demonstrating the benefits they can bring to farmers.

In this chapter, a comprehensive description of the Agroconnect research station is presented. For this aim, Sect. 3.2 provides a detailed description of the facility, along with an overview of the control problems it presents. Section 3.3 describes the benchmark design, including basic formulation of the different control problems. Section 3.4 presents the cyber-physical architecture available for accessing the benchmark. Finally, Sect. 3.5 delves into the techniques that have been applied in existing literature and provides recommendations for future research endeavors.

3.2 Problem Description

3.2.1 General Agroconnect Description

The heart of Agroconnect is an experimental greenhouse of 1900 m^2 surface area (presented inside Fig. 3.2). It is an enclosure in which climatic conditions can be regulated to favor the growth of plants. The climate inside a greenhouse is the result of the effect of different variations in the climate outside. For example, solar radiation passes through the cover and causes an increase in the temperature of the air inside the greenhouse, which can be beneficial to accelerate plant growth. However, in certain circumstances, the outdoor climate can create undesirable conditions for the indoor climate, such as high or low temperatures, which can damage the crop, so actuators are required to maintain proper conditions for crop development. Therefore, the greenhouse is equipped with a wide range of systems that encompass the water–energy–food–carbon (WEFC) nexus [2]. As described below, the auxiliary systems of the greenhouse are facilities that consume, produce, or store energy and resources. It is therefore of particular interest to study the WEFC nexus in the Agroconnect facilities through the application of automatic control and efficient management of the resources exchanged for the greenhouse operation, for instance, to achieve the desired crop production target with a reduced water and energy use.

The natural ventilation system of the greenhouse consists of thirteen motors with a power of 90 W each, six of which are used for the zenithal windows and the rest for the lateral bands. This system allows the indoor temperature and relative humidity to be maintained at the desired target level. To complement indoor temperature control, a heating and cooling system is available in the greenhouse. It consists of a 60 kW reversible heat pump with a coefficient of performance (COP) of 4, a 160 kW biomass boiler powered mainly by pellets, and a 4000 L buffer tank that exchange heat with the greenhouse using six 26.1 kW thermal power fan coils and an air flow rate of 1990 m^3/h each, which provides a homogeneous distribution of heat or cold inside the greenhouse. To enhance the regulation of indoor relative humidity, the greenhouse is equipped with a humidification system comprising 125 nebulizers located every

1.6 m over the entire surface of the greenhouse, a 7.5 hp air compressor, and a 1.5 kW electric pump, as well as a dehumidification machine of 11.2 kW.

High-tech greenhouses have demonstrated that the rate of photosynthesis, and thus crop yield, can be improved if additional CO_2 is supplemented inside the greenhouse. In Agroconnect facility, the concentration of CO_2 within the greenhouse primarily relies on two systems. Firstly, there is a CO_2 capture and injection system that employs a drip system with emitters placed every 0.25 m along the greenhouse surface. This system comprises a pure CO_2 tank of 6190 L and an activated carbon tank, which stores CO_2 from the combustion in the boiler and injects it into the greenhouse through a 5.5 kW blower. Secondly, the ventilation rate due to the exchange of air masses through the windows also plays a significant role.

To minimize natural freshwater consumption, and considering the proximity of the facility to the sea, two desalination units are available. The first unit employs reverse osmosis (RO) technology, with a specific consumption of 3 kWh/m^3 that directly draws seawater from a 100 m^3 tank. This RO unit generates 11 m^3/day of osmotized water, which is stored in a separate 100 m^3 tank. The osmotized water is utilized for both the fertigation system and the heating and cooling system. Furthermore, the RO unit produces 22 m^3/h of brine, which is stored in a 50 m^3 tank and used as a feed for the second desalination system. This second unit consists of a membrane distillation (MD) system capable of producing 6 m^3/day of distilled water. This water is stored in a tank and used to supply the humidification system. The MD requires both a heat and a cold source in order to produce distilled water. Continuous heat is provided by a solar thermal field comprising 30 flat-plate collectors. Also, a biomass boiler is available as an auxiliary source to meet the thermal demands when the power of the solar thermal system is insufficient. Finally, the sea is used as the cooling source.

Regarding electricity, two primary sources are used to ensure a reliable and sustainable electrical supply: the electrical grid and an on-grid photovoltaic generation system. The photovoltaic system consists of two units, each comprising 18 solar panels, providing a combined power generation capacity of 16.2 kW. The surplus electricity generated by the photovoltaic solar field is redirected to the grid.

3.2.2 Main Control Problems Within Agroconnect Environment

Greenhouse Control Problem

As shown in Fig. 3.2, from the point of view of automatic control, the controlled variables in greenhouses are those that directly affect crop growth, such as temperature, solar radiation, humidity, CO_2, and fertigation aspects (i.e., nutrients and water). The disturbances are the climate outside the greenhouse, the crop transpiration, plagues and diseases, and human interventions. The manipulated variables are those that compensate for the undesired effects of disturbances, such as natural ventilation, heating, humidification, and irrigation, among others.

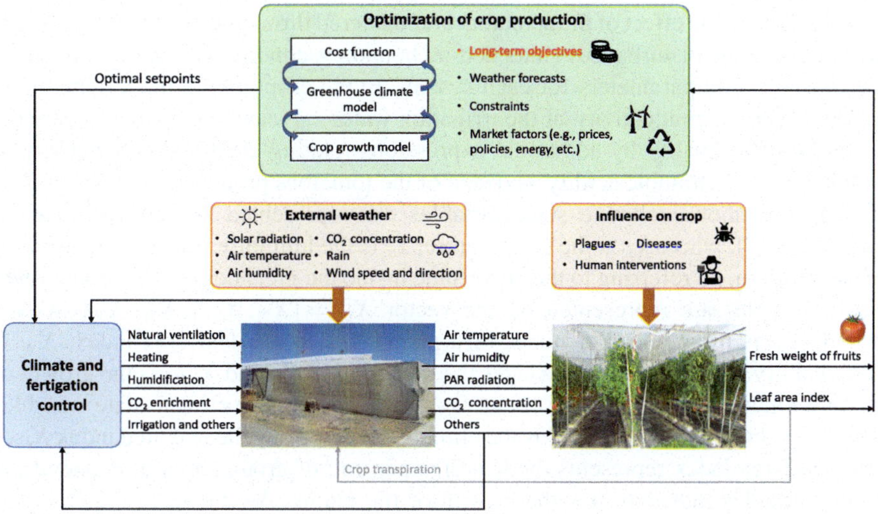

Fig. 3.2 Schematic description of the greenhouse control problem

The objective of the greenhouse climate control is to maintain optimal conditions for crop growth despite the disturbances that affect the problem. In this sense, it is important to note that the crop growth has a slow dynamic, lasting days or months, while the evolution of the greenhouse climate has a fast dynamic, measured in seconds or minutes. The climate inside the greenhouse is therefore controlled to optimize fruit production based on long-term goals, which are usually expressed in economic terms. These goals are multiple and include not only increasing production, but also maximizing economic profit, minimizing energy and water consumption required by the actuators, and improving fruit quality. As shown in Fig. 3.2 and Table 3.1, the greenhouse control problem is typically addressed by means of hierarchical control architectures to deal with the different time scales. Thus, there are two different but coupled control problems: (i) crop production optimization, which is solved in upper layers, and (ii) greenhouse climate and fertigation control at lower layers.

The relationship between greenhouse climate and crop growth can be modeled with two systems of differential equations as follows:

$$\frac{\mathrm{d}\boldsymbol{x}_{\mathrm{cl}}}{\mathrm{d}t} = f_{\mathrm{cl}}(\boldsymbol{x}_{\mathrm{cl}}, X_{\mathrm{LAI}}, \boldsymbol{u}, \boldsymbol{d}, \boldsymbol{c}_{\mathrm{cl}}, t) \tag{3.1}$$

$$\frac{\mathrm{d}\boldsymbol{X}_{\mathrm{cr}}}{\mathrm{d}t} = f_{\mathrm{cr}}(\boldsymbol{X}_{\mathrm{cr}}, \boldsymbol{x}_{\mathrm{cl}}, x_{\mathrm{EC}}, \boldsymbol{d}, \boldsymbol{c}_{\mathrm{cr}}, t) \tag{3.2}$$

where f_{cl} is usually a first-principles model based on nonlinear expressions for energy fluxes and mass balances, f_{cr} is a crop growth model based on nonlinear expressions for physiological aspects such as photosynthesis and respiration of the plants, and t is the time. For the greenhouse of Agroconnect, a tomato crop growth model is used. On the other hand, the effect of the actuators is considered through $\boldsymbol{u}$ as a vector of inputs, $\boldsymbol{d}$ is a vector of disturbances (intended as boundary conditions), and $\boldsymbol{c}$ is a vector of constants and parameters for each system. In the crop growth model, the effect of the electrical conductivity of the irrigation water x_{EC} can be treated as a variable to assess fruit quality by additional expressions relating electrical conductivity to soluble solids, titratable acidity, and size of the tomatoes produced.

The greenhouse climate state variables are represented by the vector $\boldsymbol{x}_{\mathrm{cl}} = [x_{\mathrm{T}}, x_{\mathrm{h}}, x_{\mathrm{CO2}}]$, where x_{T} is the temperature, x_{h} is the humidity, and x_{CO2} is the CO_2 concentration, all referring to the air volume inside the greenhouse. The crop growth state variables are represented by the vector $\boldsymbol{X}_{\mathrm{cr}} = [X_{\mathrm{N}}, X_{\mathrm{LAI}}, X_{\mathrm{W}}, X_{\mathrm{WF}}, X_{\mathrm{WM}}]$, where X_{N} is the number of nodes of a plant, X_{LAI} is the leaf area index, X_{W} is the total dry weight of a plant, X_{WF} is the fruit dry matter, and X_{WM} is the growth of mature tomatoes. Note that the vector $\boldsymbol{x}_{\mathrm{cl}}$ of the greenhouse climate state variables affects the crop growth, as well as the climate is affected by the leaf area index X_{LAI}. The leaf area index represents the leaf area per unit of ground area. It is one of the main variables that indicates the growth of the plants. As shown in Fig. 3.2, it is important to take into account the effect of this variable due to crop transpiration through plants' stomata, which affects greenhouse temperature and humidity. For a complete description of the models, the reader is referred to [14].

These models are used for simulation purposes, although they can also be simplified and expressed in steady state for the crop production optimization problem. On the other hand, it can be complicated to design low-level control strategies for greenhouse climate control using these models, so data-driven system identification techniques are typically used to obtain linear auto-regressive models and/or first-order transfer functions with dead time [4].

There are several important factors to consider when implementing optimization and control strategies for greenhouses, but the most relevant are as follows: (i) The calculation of the optimal setpoints must distinguish between day and night, because the weather conditions are different and the plants photosynthesize during the day, when they require higher temperature setpoints. At night, the temperature setpoints can be lower because the crop does not grow. (ii) During the day, the effect of disturbances is strong, so the problem of greenhouse climate control consists of both tracking and regulation. (iii) Some variables that have a major impact on plant growth and health, such as temperature and humidity, are inversely correlated. (iv) Saturation of the actuators can affect control performance.

Desalination System Control Problem

In the desalination systems, two different control problems can be distinguished. For the RO system, the control problem consists of a scheduling problem, which is generally treated from the point of view of energy hubs (see Sect. 3.2.2). In the MD system, the control problem (see Fig. 3.3) is more complex due to the nature of the system and the different disturbances involved [8]. Thus, the use of adequate low-level control strategies is required to achieve the desired system's performance.

As can be seen in Fig. 3.3, the system's manipulated variable is the module feed flow rate, while the primary disturbances are linked to the feed water conditions (salinity and temperature) and the operating temperature at the evaporation channel inlet, typically influenced by the performance of a solar thermal field [8]. As commented, the major objective pursued in this case is to maintain a certain level of distillate production, according to the demand of the greenhouse. An additional objective involves minimizing the system's specific thermal energy consumption (STEC), which is related to the thermal consumption of the system, and, therefore, to the operating costs [9]. Both the distillate production and STEC, for a specific discrete instant time k, are given by

$$\mathrm{D}^k = f_D(u^k, \mathbf{d}^k), \qquad \mathrm{STEC}^k = f_{STEC}(D^k, u^k, \mathbf{d}^k), \tag{3.3}$$

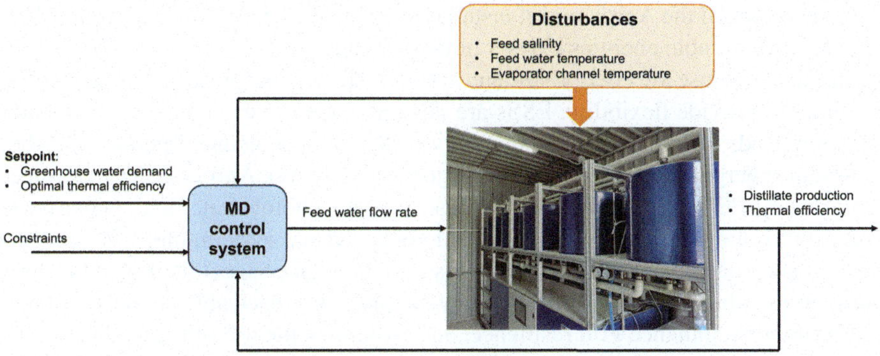

Fig. 3.3 Schematic description of the MD control problem

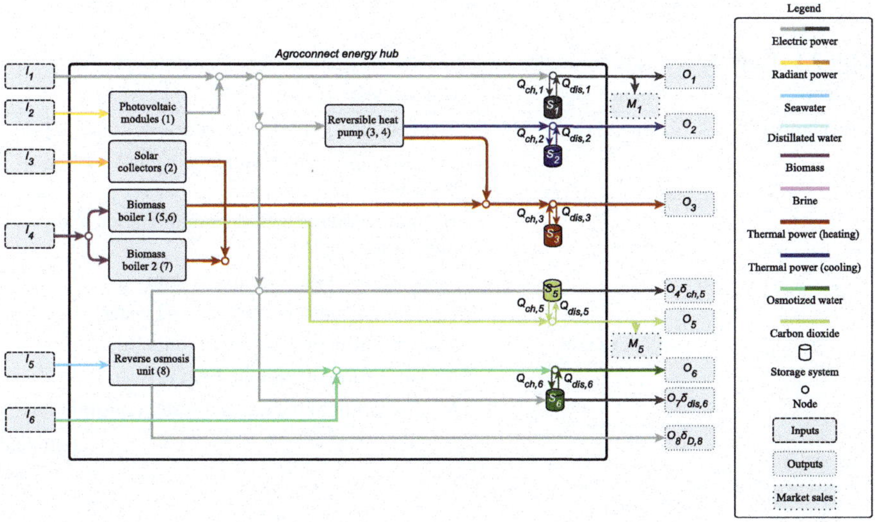

Fig. 3.4 Agroconnect energy hub model

where f_D and f_{STEC} are nonlinear static functions while the rest of the variables follows the previous introduced nomenclature. The central challenge of this process from a control point of view is that both functions are difficult to obtain due to the system's nonlinear response to changes in salinity [8]. In addition, the system is subjected to constraints at the input of the process and, depending on the objective established in the control problem, at the output [9].

Agroconnect Energy Hub

For the efficient, sustainable, and autonomous management of Agroconnect facility, energy costs must also be taken into account through an economic study. In this context, the EH concept arises as a model that provides the relationships between inputs, outputs, conversions, and storage of multiple energy carriers [7]. Figure 3.4 shows the EH scheme of the Agroconnect installation using the energy hub paradigm where the use of renewable energies plays a fundamental role.

In EHs, the use of energy storage systems (ESSs) becomes an indispensable resource to provide flexibility. ESSs are designed to store excess energy generated during periods of low demand and supply it during peak demand periods or when renewable energy sources are not actively generating power. Agroconnect is equipped with several ESSs as explained before: batteries for photovoltaic field, water reservoirs for cooling and heating system and solar thermal system, and CO_2 tank for greenhouse injection. By integrating ESSs into the Agroconnect facility, it becomes possible to balance the fluctuating energy supply from renewable sources, mitigate grid instability, enhance grid resilience, and maximize the use of renewable energy. Additionally, ESSs contribute to a more sustainable energy ecosystem by minimizing wastage and reducing reliance on fossil fuel-based backup power generation.

Table 3.1 Examples of possible control tests that can be performed in Agroconnect

System	Control layer	Control test
Greenhouse	Lower layer	Temperature control with natural ventilation [4]
		Temperature control with heating and cooling systems
		Humidity control with dehumidification and humidification [5]
		CO_2 control with injection system
		Fertigation control (nutrients and water)
	Upper layer	Maximization of economic profits [6]
		Maximization of fruit quality [12]
		Minimization of energy and water consumption

(continued)

Table 3.1 (continued)

System	Control layer	Control test
Energy hub	Unique layer	Overall system minimization of economic costs [13]
		Overall system minimization of non-renewable energies consumption
		Overall system minimization of water consumption
MD unit	Lower layer	Distillate production control
		Thermal efficiency control [9]
		Temperature control of the thermal power source (solar field) [8]
		Temperature control of the refrigeration circuit (based on sea water)
	Upper layer	Minimize thermal energy consumption [8]
		Maximize distillate production [8]
		Minimize the levelized cost of water

3.3 Benchmark Design

The main benchmark proposed with Agroconnect is to optimize and coordinate the operation of all the experimental facilities described in Sect. 3.2 to ensure a sustainable crop production in the greenhouse. This benchmark can be attractive for the field of control engineering due to the number of systems involved with different dynamics and time scales, allowing to test a wide variety of control techniques. In this context, as shown in Fig. 3.1, the main challenge is to combine the optimization of crop production with an optimal resource exchange between facilities. The specific control problems that can be performed in Agroconnect are summarized in Table 3.1.

Particularizing in the greenhouse, the main objective is to maximize profits from crop production, based on the following cost function:

$$J(x_{\mathrm{T}}, x_{\mathrm{EC}}) = \int_{t_{\mathrm{i}}}^{t_{\mathrm{f}}} c_{\mathrm{area}}\, P(t)\, X_F(t)\, \mathrm{d}t - \int_{t_{\mathrm{i}}}^{t_{\mathrm{f}}} V_{\mathrm{cost}}(t)\, \mathrm{d}t \tag{3.4}$$

where t_{i} is the start of the crop cycle, t_{f} is the time that the last fruits are harvested, c_{area} is the greenhouse soil area, P is the market price of the fruits, and X_{F} is the fresh weight of mature fruits (calculated from X_{WM}). In this example, J is a performance metric of the crop production optimization problem, which is considered to depend on the greenhouse temperature x_{T} and the electrical conductivity of the irrigation water x_{EC}. These variables have an impact on crop growth according to Eqs. (3.1) and (3.2), so constraints should be imposed to ensure that setpoint values are feasible and within appropriate ranges depending on the type of crop and the weather conditions in each season. To achieve the optimal setpoints for temperature and electrical conductivity, the actuators used to control the greenhouse climate consume energy and resources (i.e., electricity, heat, water, and CO_2), which is reflected in the costs term $V_{\text{cost}}(t)$.

Moreover, optimal resource management is required to meet the resource needs of the greenhouse, but also to consider the expected consumption or production of resources by other facilities (i.e., electricity production by photovoltaic panels). Following the EH paradigm, the optimization of resources can be expressed as

$$\min \sum_{k=1}^{H} (\mathbf{P}_{\text{r}}(k)\, \mathbf{I}_{\text{f}}(k) - \mathbf{S}_{\text{r}}(k)\, \mathbf{M}_{\text{f}}(k)) \tag{3.5}$$

where $\mathbf{P}_{\text{r}}(k)$ is a vector containing the price for resources purchased, $\mathbf{I}_{\text{f}}(k)$ is a vector for the resource input flows, $\mathbf{S}_{\text{r}}(k)$ is a vector containing the price of the resources sold, $\mathbf{M}_{\text{f}}(k)$ is a vector for market-sales flows, H is the prediction horizon, and k are the discrete-time instants. This optimization problem is subject to a series of constraints concerning the physical capacities of the systems, such as operational flows and/or storage limits.

Different scenarios should be considered for the two optimization problems depending on market and energy prices, weather forecasts, environmental criteria, or water saving policies. Moreover, it is important to highlight the need to know the state of the crop reliably every day and to have accurate models of resource consumption and/or production for each system.

In addition to the optimization problems mentioned above, the greenhouse climate control itself is another benchmark (see Table 3.1), for which the reliability of the control techniques should be guaranteed, ensuring that the greenhouse climate can be controlled at the optimal setpoints for both day and night, and in all seasons. The integral of the absolute error is often used to evaluate low-level control performance, primarily for disturbance rejection.

Finally, another interesting benchmark for control engineers is related to the MD unit. In this case, different problems can be distinguished as well, spanning from low-level control problems concerning the regulation of operating temperatures through the solar field to control challenges associated with water cost considerations; see Table 3.1. Furthermore, special attention is warranted for aspects related to the performance of the MD unit itself, such as distillate production and thermal efficiency control (see Table 3.1). These aspects pose a challenge for system modeling, thus making the case for the use of model-free controllers or data-driven strategies [9]. For the application of these controllers, stationary measurements of distillate production are available in the real system, as well as measurements of the main disturbances affecting the process. Therefore, the controllers must be focused on conducting the systems to an optimal steady-state operating point by iteratively adapting the process input. This optimization problem can be mathematically expressed as

$$\min_{u} \; J(D) \tag{3.6}$$

$$\text{s.t.} \;\; \mathbf{A}u \leq \mathbf{B}, \tag{3.7}$$

where J is a continuously differentiable objective function aimed at maintaining the desired level of distillate production, minimizing STEC, or even considering a multi-objective approach to contemplate both objectives simultaneously. **A** and **B** are constant matrices related to the input constraints of the problem.

3.4 Data Access Architecture

Interoperability and data accessibility are essential elements to ensure reliable control operations in a plant. In the case of the Agroconnect, which has a diversity of devices and heterogeneous data sources, the proper integration of this data becomes critical. What is more, without proper interoperability and ease of data extraction, the ability to perform effective plant control is greatly hindered.

The Agroconnect infrastructure is composed of a diversity of devices that generate a heterogeneous set of data (see Sect. 3.2). These data sources include IoT stations, desalination systems, isolated sensors and actuators, irrigation systems, and photovoltaic panels, among others. Each system uses its own communication protocol, making it difficult to extract and process data to apply control techniques in the plant.

To achieve efficient integration, the FIWARE OMA NGSI data model is used, which provides a standard in data representation [1]. FIWARE is a platform developed by the European Union for the IoT paradigm, with the aim of establishing the necessary interoperability in the IoT sector. In addition, a set of OPC UA servers is used as an intermediate interface to enable bidirectional communication between the cloud and edge devices. This configuration makes it possible to subscribe clients to the OPC UA servers, enabling control operations in the cloud and providing flexibility for companies or research centers to operate on a specific part of the plant without compromising the security of data on other servers.

Data extraction is performed through a modular architecture using data extraction, transformation and loading (ETL) systems, and periodic command run on (CRON) processes [11].

The Agroconnect modular architecture integrates all available services in different layers (see Fig. 3.5). The edge layer is located at the bottom and houses the sensors, actuators, and IoT devices, enabling local control with low latency and avoiding communication problems in critical systems. The middle layer, known as the fog layer, is composed of microclouds [15] containing components such as weather measurements, power production, water facilities, and crop growth variables. This fog layer provides separate OPC UA servers per service, allowing clients to adhere to a specific fog and perform control operations without affecting the overall architecture and other fogs. At the top layer, there are services exposed to users, such as control panels, REST API services for information extraction, and cloud models. In this layer, the standardized FIWARE OMA NGSI data model is used and ETL systems are used to extract data from the lower layers and transform it to the upper layer.

The Agroconnect infrastructure provides stakeholders with access to benchmarking data and resources through a REST API. This allows users to make HTTP requests

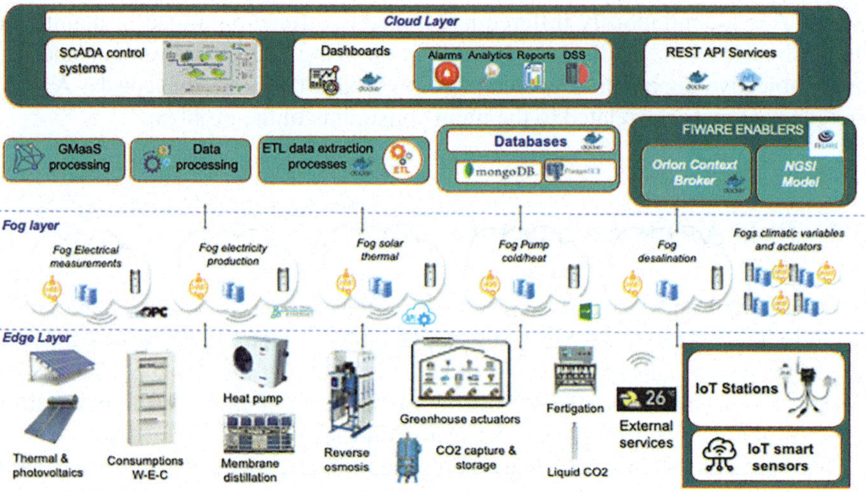

Fig. 3.5 IoT architecture for data integration in Agroconnect

from any REST client, facilitating integration with other systems and applications. There are no limitations or licensing requirements associated with accessing benchmarking data or resources, as it is an open-access cloud service. No specific tools or software packages are required to use the benchmark. The data is accessible from any computer with an Internet connection, providing flexibility and convenience to users. The Agroconnect infrastructure also offers a Swagger-based DOC API, making the REST API easy to understand and use. Documentation is standardized and provides clear guidance on how to interact with the available data and resources. Remote participation is made possible by the availability of data in the cloud. Users can perform modeling in the cloud and send data to the systems to perform a control benchmark. This allows stakeholders to participate and use the Agroconnect infrastructure without the need to be physically present at the plant.

3.5 Discussion for Future Comparison

As explained in Sect. 3.2, the problem of controlling greenhouse crop production is usually addressed by means of hierarchical control architectures. MPC and optimal control methods are used in the upper layers, while a wide variety of well-known control techniques are used for climate control in the lower layers, including PID-based controllers [14]. For instance, Fig. 3.6 presents real results of temperature control in the Agroconnect greenhouse after testing a PI controller designed for setpoint tracking and disturbance rejection. As observed, the test consisted of regulating the temperature inside the greenhouse by adjusting the opening of the vents.

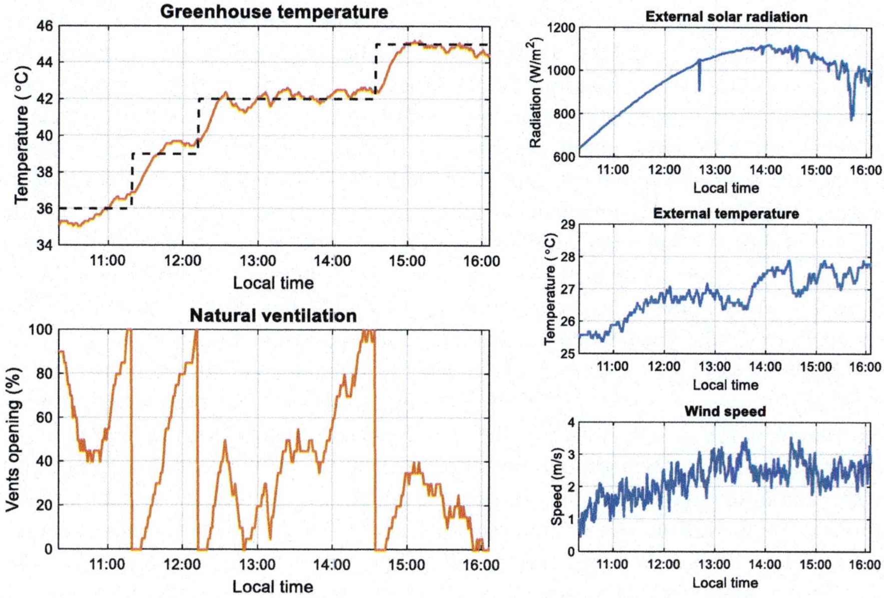

Fig. 3.6 Example of temperature control in the Agroconnect greenhouse

In the upper layers, crop production optimization can be approached as a multi-objective optimization problem, as shown in [12]. The main strength of this approach is that the optimal setpoints for the greenhouse air temperature and for the electrical conductivity of the irrigation water are calculated according to a compromise solution, selected from a Pareto front which combines the objectives of maximization of profits, quality, and water-use efficiency. The main weakness is that the existing approaches do not take into account the uncertainty of important factors such as weather forecasts or market prices, which are essential for long-term objectives. Recently, a stochastic control technique, based on multi-scenario model predictive control, has been developed in [6] to account for market price uncertainty and maximize the profit that farmers receive from selling the crop production. Although the weakness is that the optimal solution could be over-conservative, the main strength is that possible realizations of market prices can be considered in the optimization problem and economic risks are minimized by selecting the optimal dates to harvest and sale the mature fruits.

In summary, in the upper layers of hierarchical control architectures, the current research gaps are related to solving the coupled problems of crop production optimization and resource scheduling. This implies the need to coordinate the operation of the different facilities of an agro-industrial district, imposing environmental and sustainability criteria. In this context, approaches such as distributed control, coalitional control, and game theory [3] could be studied to consider the different facilities as multiple agents with different needs but also with a common overall objective.

Regarding low-level control focused on greenhouse climate control, temperature, radiation, humidity, and CO_2 are the main variables to regulate, in addition to fertigation aspects. When the required actuators are available in a greenhouse, such as natural ventilation, heating, shading mesh, humidification and dehumidification systems, and CO_2 enrichment systems, the adequate regulation of these variables has been demonstrated to be possible with both classical and advanced control techniques [14]. However, the gaps in current research are in how to effectively combine the use of these multiple actuators to keep each variable at its desired setpoint while avoiding the effects of disturbances and interactions between control loops. Multivariable control approaches are a possible solution [5], but more research is needed, specially to validate these techniques under real crop production conditions.

Moreover, in order to achieve an efficient, sustainable, and autonomous management of resources in Agroconnect, the paradigm of the EHs has been used following the scheme in Fig. 3.4. Specifically, the Agroconnect infrastructure has been modeled in accordance with the methodology proposed by [13]. This modeling framework offers numerous advantages, allowing for the integration of production, consumption, and storage models across its subsystems. This enables efficient resource management and facilitates decision-making over short, medium, and long-term periods. However, certain challenges arise due to the extensive use of renewable energies, the fluctuating prices of resources, and the escalating frequency of weather events. These factors significantly impact the reliability of resource planning.

To tackle the above-mentioned challenges and ensure robust decision-making, it becomes imperative to adopt methodologies that embrace uncertainties associated with the elements involved. An interesting illustration of this is found in the proposal for a risk-averse scheduling approach for an electricity-heat system outlined in [10]. This exemplifies the need to account for uncertainties in various aspects of the system when making decisions. These types of methodologies have not been applied to Agroconnect facilities, where there is a wide variety of resources in an agro-industrial environment. Looking ahead, as a part of future research, the goal is to adapt the existing EH methodology applied in Agroconnect to incorporate the effects of multiple uncertainties present in the models. The incorporation of uncertainty considerations in the planning process is a relevant aspect, as it provides a more realistic and comprehensive representation of the system behavior. This comprehensive approach fosters a sustainable and adaptable Agroconnect infrastructure, capable of adapting to dynamic and uncertain scenarios. In conclusion, the capacity to model each subsystem, encompassing production, consumption, and storage, empowers efficient resource management and strategic decision-making. Nevertheless, the challenges posed by renewable energy usage, fluctuating resource prices, and weather variability require the incorporation of uncertainty-aware methodologies.

Finally, another fundamental challenge to be addressed involves the real-time optimization of MD module operations. However, the complexity of modeling these systems presents a significant obstacle in the development and application of effective model-based controllers. Consequently, model-free approaches have emerged as an attractive and widely adopted choice within the MD field. Notably, various Extremum Seeking Control (ESC) and Online Feedback Optimization (OFO) strategies have been proposed in the existing literature [9]. Despite the demonstrated effectiveness of these techniques, they exhibit certain limitations, notably slow convergence to optimal solutions and vulnerability to noise when the system is perturbed. Consequently, future research efforts in this field should be concentrated on addressing and improving these aspects.

Acknowledgements This work is a result of the CyberGreen Project, PID2021-122560OB-I00, and the Agroconnect facilities (www.agroconnect.es), grant EQC2019-006658-P, both funded by MCIN/AEI/ 10.13039/501100011033 and by ERDF A way to make Europe.

References

1. Cirillo F, Solmaz G, Berz EL, Bauer M, Cheng B, Kovacs E (2019) A standard-based open source IoT platform: Fiware. IEEE Internet Things Mag 2(3):12–18
2. Cui S, Zhang J, Wang X, Wu M, Cao X (2023) Fuzzy composite risk assessment of water-energy-food-carbon nexus in the dispark pumped irrigation system. J Hydrol 624:129879
3. Fele F, Maestre JM, Camacho EF (2017) Coalitional control: Cooperative game theory and control. IEEE Control Syst Mag 37(1)
4. García-Mañas F, Guzmán JL, Rodríguez F, Berenguel M, Hägglund T (2021) Experimental evaluation of feedforward tuning rules. Control Eng Practice 114:104877
5. García-Mañas F, Hägglund T, Guzmán JL, Rodríguez F, Berenguel M (2024) A practical solution for multivariable control of temperature and humidity in greenhouses. Euro J Control 77:100967
6. García-Mañas F, Rodríguez F, Berenguel M, Maestre JM (2024) Multi-scenario model predictive control for greenhouse crop production considering market price uncertainty. IEEE Trans Automat Sci Eng 21(3):2936–2948
7. Geidl M (2007) Integrated modeling and optimization of multi-carrier energy systems. PhD thesis, ETH Zurich
8. Gil JD (2023) *Hierarchical control and optimization strategies applied to solar membrane distillation facilities*. Springer Nature
9. Gil JD, Bueso A, Roca L, Zaragoza G, Berenguel M (2023) Data-driven online feedback optimization of solar membrane distillation systems operating in batch mode. J Process Control 129:103056
10. Lu X, Li H, Zhou K, Yang S (2023) Optimal load dispatch of energy hub considering uncertainties of renewable energy and demand response. Energy 262:125564
11. Muñoz M, Guzmán JL, Sánchez-Molina JA, Rodríguez F, Torres M, Berenguel M (2022) A new IoT-based platform for greenhouse crop production. IEEE Internet Things J 9(9):6325–6334
12. Ramírez-Arias A, Rodríguez F, Guzmán JL, Berenguel M (2012) Multiobjective hierarchical control architecture for greenhouse crop growth. Automatica 48(3):490–498
13. Ramos-Teodoro J, Giménez-Miralles A, Rodríguez F, Berenguel M (2020) A flexible tool for modeling and optimal dispatch of resources in agri-energy hubs. Sustainability 12:1–24
14. Rodríguez F, Berenguel M, Guzmán JL (2015) Modeling and control of greenhouse crop growth. Springer, Cham, Switzerland
15. Rolik O, Zharikov E, Telenyk S (2017) Microcloud-based architecture of management system for IoT infrastructures. In: 2016 3rd international scientific-practical conference problems of infocommunications science and technology, PIC S and T 2016—Proceedings, pp 149–151

Chapter 4
Temperature Control of a Shrink Tunnel with Multiple Heating Zones

Davide Previtali, Leandro Pitturelli, Fabio Previdi, and Antonio Ferramosca

4.1 Introduction

Shrink tunnels are widely used in manufacturing applications for polymeric packaging, such as bottle packs manufacturing processes. The system under study, depicted in Fig. 4.1, is composed of an *industrial oven* and a *conveyor belt* that feeds bottled beverages to it. The industrial oven is 4.10 m long, 2.20 m tall, and 1.40 m wide; instead, the conveyor belt is 5.70 m long. The speed of the conveyor belt can vary based on the production rate of the industrial process which produces the beverages supplied to the shrink tunnel. Before being inserted into the oven cavity, the bottles are mechanically grouped together and covered by a thin plastic film. An *infrared sensor* placed at the entrance of the oven detects when the beverages are being fed to it (see Fig. 4.1a). The heat shrinks the plastic and tightly envelopes the bottles, creating the pack. The oven cavity can house several bottles simultaneously, due to its large volume of 2.26 m^3, and is divided into two interconnected *heating zones*. The heat in each zone is produced by a dedicated set of *heat resistors*, located in separate compartments (see Fig. 4.1b). In order to tightly shrink the plastic film around the bottles there must be a suitable temperature profile within the cavity. Four *convection fans* allow for air circulation inside the oven and serve two purposes: (i)

D. Previtali (✉) · L. Pitturelli · F. Previdi · A. Ferramosca
Department of Management, Information and Production Engineering,
University of Bergamo, Via G. Marconi 5, 24044 Dalmine, BG, Italy
e-mail: davide.previtali@unibg.it

L. Pitturelli
e-mail: leandro.pitturelli@unibg.it

F. Previdi
e-mail: fabio.previdi@unibg.it

A. Ferramosca
e-mail: antonio.ferramosca@unibg.it

J. M. Maestre and C. Ocampo-Martinez (eds.), *Control Systems Benchmarks*, Advances in Industrial Control, https://doi.org/10.1007/978-3-031-76312-0_4

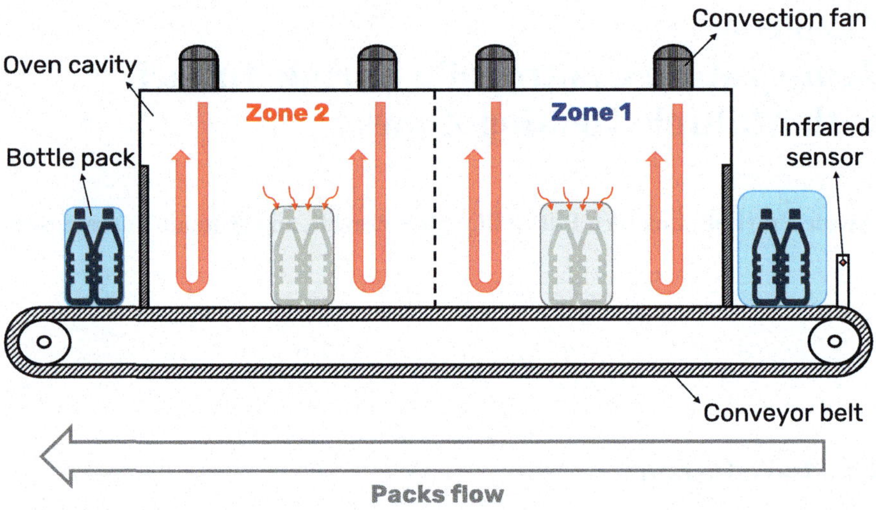

(a) Front view.

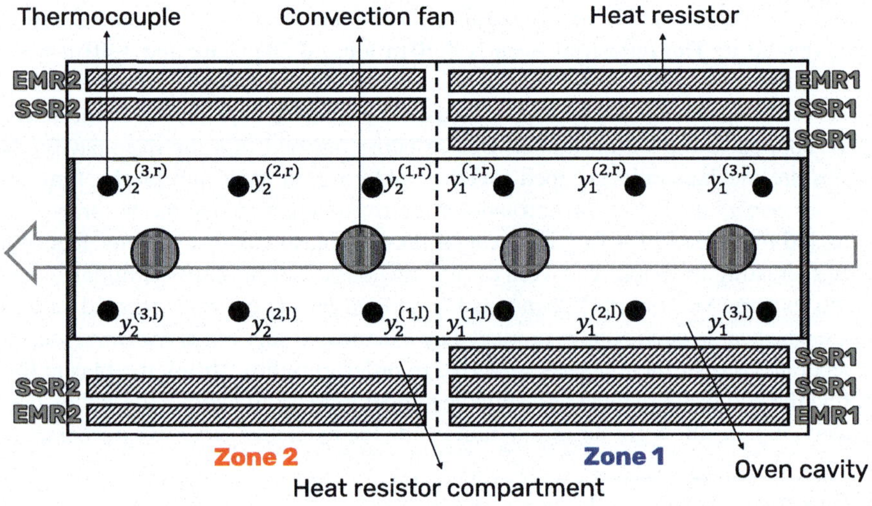

(b) Top view.

Fig. 4.1 Schematic of the considered shrink tunnel. $y_i^{(l,s)}$, $i \in \{1, 2\}$, $l \in \{1, 2, 3\}$, $s \in \{\text{l}, \text{r}\}$, denote the temperatures measured by the twelve thermocouples. The heat resistors in zone i denoted by EMRi are driven by the same electromechanical relay while those denoted by SSRi are driven by the same solid-state relay convey hot air coming from the heat resistors to the cavity through several openings in the compartments, and (ii) ensure that the plastic film, when shrunk, covers the bottles uniformly. The air temperature inside the oven cavity is measured by twelve approximately equally spaced *thermocouples* (six per zone). Each sensor is located at the same height (0.4 m) from the conveyor belt and spaced about 0.5 m apart from the adjacent thermocouples. In Fig. 4.1b, we denote the temperature measured by each sensor as $y_i^{(l,s)} \in \mathbb{R}$ (in °C) where

- $i \in \{1, 2\}$ denotes the *zone* of belonging (either the first or the second one),
- $l \in \{1, 2, 3\}$ denotes the *location* associated with the i-th zone. In particular, $l = 1$ is associated with those thermocouples that are close to the center of the oven, while the temperatures with $l = 3$ are at the openings of the oven cavity,
- $s \in \{\text{l}, \text{r}\}$ indicates whether the thermocouple is located on the left (l) or the right (r) *side* of the cavity.

The heat resistors in each zone are controlled either in groups of two or four. In particular, each zone relies on a *Solid-State Relay (SSR)* and an *ElectroMechanical Relay (EMR)* to control the voltages across the heat resistors via *Pulse-Width Modulation (PWM)*. Figure 4.1b highlights which heating elements are driven by the EMRs and which ones by the SSRs. The relays are responsible for producing the voltage PWM signals based on the duty cycles supplied by a temperature controller. Using different types of relays adds an additional layer of complexity to the system: EMRs typically exhibit slower response times and must be operated at lower switching rates compared to SSRs to prevent degradation [2]. For this reason, the periods of the PWM signals are set to 30 s for the EMRs and 1 s for the SSRs. However, EMRs are usually cheaper than SSRs, mitigating the overall production costs of the shrink tunnel. Consequently, the combination of these two types of relays leads to reasonable response times at an affordable price.

The air temperature inside the oven cavity must be controlled in a closed loop to ensure correct operation of the shrink tunnel since it directly affects the quality of the bottle pack wrapping. In particular, the *control objectives* are: (i) keep the temperature values measured by the twelve thermocouples within a tight working range, creating a suitable temperature profile in the oven cavity, along the direction of movement of the bottle packs; (ii) disturbance rejection, i.e., minimize the temperature drop due to the insertion of the packs into the oven; (iii) limit the energy consumption of the heat resistors. Apart from the demanding control objectives, designing a control strategy for the considered shrink tunnel can be challenging for several other reasons. First, the heat resistors in each zone are forced to operate at (roughly) the same duty cycle, limiting the effective control actions to two (one duty cycle per zone) and making the system *underactuated* [13]. Second, the system is nonlinear: the heat generated by the heat resistors depends on the square of the grid voltage. Third, the system exhibits time delays. Lastly, the heat resistors are designed to operate at a given nominal voltage, which might differ from the grid voltage. Thus, the duty cycles of their corresponding PWM signals must be computed accordingly.

4.2 Problem Description

In what follows, in Sect. 4.2.1, we derive a *lumped-parameter state-space model* for the shrink tunnel under study. To ease the derivation of the just-mentioned model and favor control design, we make the following simplifying assumption: the temperature at each spot in Fig. 4.1b results from the sum of two contributions. The former is due to the heat generated by the heat resistors and the ambient temperature, whose relationship is described by the `Heat-Temperatures` model in Sect. 4.2.1. The latter models the temperature drop caused by the insertion of bottle packs inside the oven cavity (see the `Packs-Temperatures` model in Sect. 4.2.1). Specifically, at each time $t \in \mathbb{R}_{\geq 0}$ (in s), the temperatures $y_i^{(l,s)}(t)$ in Fig. 4.1b amount to

$$y_i^{(l,s)}(t) = y_{T_i}^{(l,s)}(t) + y_{d_i}^{(l,s)}(t), \quad \forall i \in \{1, 2\}, l \in \{1, 2, 3\}, s \in \{\mathrm{l}, \mathrm{r}\}, \tag{4.1}$$

where $y_{T_i}^{(l,s)}(t) \in \mathbb{R}$ (in °C) and $y_{d_i}^{(l,s)}(t) \in \mathbb{R}$ (in °C) denote, respectively, the outputs of the `Heat-Temperatures` and `Packs-Temperatures` models. The just-described model for the system under study is summarized in Fig. 4.2, along with the temperature control architecture. Each signal in Fig. 4.2 is covered in greater detail in the next sections. Concerning the control architecture, in Sect. 4.2.2, we define the *control objectives* for the considered shrink tunnel and the *constraints on the control actions* that ensure correct and optimal operation.

4.2.1 Shrink Tunnel Modeling

Definition of the `Heat-Temperatures` Model

We describe the relationship between the heat produced by the heat resistors and the temperature at each spot in Fig. 4.1b using the *electro-equivalent thermal circuit* presented in Fig. 4.3. Electro-equivalent thermal circuits are widely used for thermal systems modeling due to their simplicity and reasonable accuracy (see, e.g., [6, 10, 11]). In this case study, we consider the oven cavity divided into twelve subzones of the same volume, each with its own thermocouple. Each subzone is modeled by an RC-analogue circuit; these are connected by a resistor that models the interaction between subzones. Specifically, let $i \in \{1, 2\}, l \in \{1, 2, 3\}$, $l' \in \{1, 2\}$, and $s \in \{\mathrm{l}, \mathrm{r}\}$. Then, referring to Fig. 4.3:

- $T_i^{(l,s)}(t) \in \mathbb{R}$ (in °C) is the temperature in zone i, location l, side s, resulting from the heat produced by the heat resistors and the ambient temperature, ignoring the contribution of the packs flow;
- $q_i(t) \in \mathbb{R}_{\geq 0}$ (in $\frac{\mathrm{J}}{\mathrm{s}}$) is the heat flow rate produced by the heat resistors located in zone i. We assume that the heat generated in each zone is equally spread among each of its locations/sides;

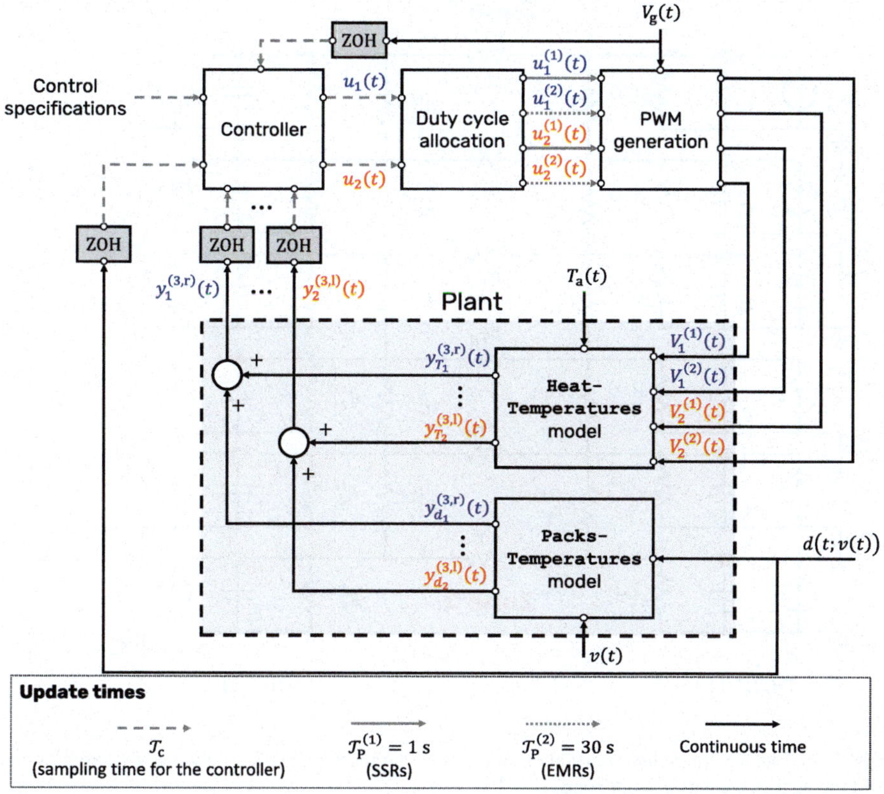

Fig. 4.2 Block diagram of the system under study. The signals in blue are associated with the first zone while those in red are related to the second zone. Instead, the signals in black are not related to any zone in particular. The different arrows highlight how often each signal is updated. Finally, the Zero-Order Holder (ZOH) blocks sample the relevant signals at a sampling time $\mathcal{T}_c$

- $T_a(t) = \bar{T}_a$, $\forall t \in \mathbb{R}_{\geq 0}$, $\bar{T}_a \in \mathbb{R}$ (in °C), is the ambient temperature. It is constant during normal operation of the shrink tunnel;
- $R_i^{(l,s)} \in \mathbb{R}_{>0}$ (in $\frac{°C}{J} \cdot s$) is the thermal resistance between the oven walls of zone i, location l, side s, and the ambient;
- $R_{T_i}^{(l',s)} \in \mathbb{R}_{>0}$ (in $\frac{°C}{J} \cdot s$) is the thermal resistance related to the transfer of heat between two locations of zone i, side s;
- $R_{T_{12}}^{(s)} \in \mathbb{R}_{>0}$ (in $\frac{°C}{J} \cdot s$) is the thermal resistance related to the transfer of heat between the two zones of the oven, for what concerns side s;
- $R_{T_{rl}} \in \mathbb{R}_{>0}$ (in $\frac{°C}{J} \cdot s$) is the thermal resistance related to the transfer of heat between the left and the right side of the oven cavity. For the sake of simplicity, we assume it to be equal in each zone and each location;
- $C_z \in \mathbb{R}_{>0}$ (in $\frac{J}{°C}$) is the thermal capacitance of the air in each zone/location/side of the oven cavity, which all have the same volume.

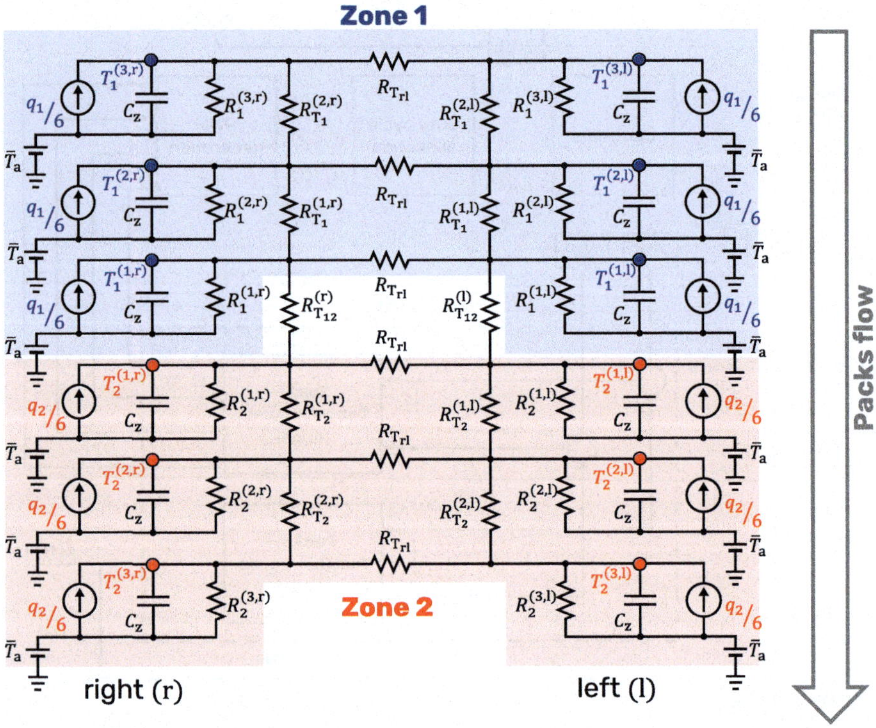

Fig. 4.3 Electro-equivalent thermal circuit describing the relationship between the temperatures and the heat generated in each zone

The relationship between each temperature $T_i^{(l,s)}(t)$ and the heat flow rates $q_i(t)$ can be readily derived from the circuit in Fig. 4.3, leading to the following system of linear differential equations [11]:

$$\dot{\boldsymbol{T}}(t) = A_{TT} \cdot \boldsymbol{T}(t) + B_q \cdot \boldsymbol{q}(t) + \boldsymbol{b}_T \cdot T_{\mathrm{a}}(t), \tag{4.2}$$

where $\boldsymbol{q}(t) = \left[q_1(t)\ q_2(t)\right]^{\top} \in \mathbb{R}^2_{\geq 0}$, while $\boldsymbol{T}(t) \in \mathbb{R}^{12}$,

$$\boldsymbol{T}(t) = \left[T_1^{(3,\mathrm{r})}(t)\ T_1^{(3,\mathrm{l})}(t)\ T_1^{(2,\mathrm{r})}(t)\ \ldots\ T_2^{(2,\mathrm{l})}(t)\ T_2^{(3,\mathrm{r})}(t)\ T_2^{(3,\mathrm{l})}(t)\right]^{\top} \tag{4.3}$$

is the vector of temperatures, going from entrance to exit and alternating between right and left. The matrices and vectors $A_{TT} \in \mathbb{R}^{12\times 12}$, $B_q \in \mathbb{R}^{12\times 2}$, and $\boldsymbol{b}_T \in \mathbb{R}^{12}$ can be computed directly from the circuit in Fig. 4.3 and are reported in the **Supplementary Material** [5]. As an example, here we report the differential equation for $T_1^{(3,\mathrm{r})}(t)$:

$$\dot{T}_1^{(3,\mathrm{r})}(t) = -\frac{1}{C_\mathrm{z}} \cdot \left(\frac{1}{R_1^{(3,\mathrm{r})}} + \frac{1}{R_{\mathrm{T}_1}^{(2,\mathrm{r})}} + \frac{1}{R_{\mathrm{T}_{\mathrm{rl}}}} \right) \cdot T_1^{(3,\mathrm{r})}(t) + \frac{1}{C_\mathrm{z} \cdot R_{\mathrm{T}_{\mathrm{rl}}}} \cdot T_1^{(3,\mathrm{l})}(t) + \\ + \frac{1}{C_\mathrm{z} \cdot R_{\mathrm{T}_1^{(2,\mathrm{r})}}} \cdot T_1^{(2,\mathrm{r})}(t) + \frac{1}{6 \cdot C_\mathrm{z}} \cdot q_1(t) + \frac{1}{C_\mathrm{z} \cdot R_1^{(3,\mathrm{r})}} \cdot T_\mathrm{a}(t).$$

To account for the *delay introduced by the thermocouples* when measuring the temperature at each spot in Fig. 4.1b, we consider a time lag $L_{T_i}^{(l,s)} \in \mathbb{R}_{\geq 0}$ (in s) for each $T_i^{(l,s)}(t)$, $i \in \{1, 2\}$, $l \in \{1, 2, 3\}$, $s \in \{\mathrm{l}, \mathrm{r}\}$. Specifically, referring to $y_{T_i}^{(l,s)}(t)$ in (4.1), we have

$$y_{T_i}^{(l,s)}(t) = T_i^{(l,s)}\left(t - L_{T_i}^{(l,s)}\right). \tag{4.4}$$

Next, we model the *propagation of heat* from the heat resistors to the air inside the oven cavity using a first-order low-pass filter with unitary gain:

$$\dot{q}_i^{(\mathrm{f})}(t) = -\frac{1}{\tau_i^{(\mathrm{f})}} \cdot q_i^{(\mathrm{f})}(t) + \frac{1}{\tau_i^{(\mathrm{f})}} \cdot q_i(t), \quad i \in \{1, 2\}, \tag{4.5}$$

where $q_i^{(\mathrm{f})}(t)$ is the filtered heat flow rate signal and $\tau_i^{(\mathrm{f})} \in \mathbb{R}_{>0}$ (in s) is the time constant for zone i. Then, in Fig. 4.3, we should replace $q_i(t)$ with $q_i^{(\mathrm{f})}(t)$.

The heat flow rates $q_i(t)$ in (4.5) are generated by the heat resistors located in the corresponding zones. We consider the heat resistors in Fig. 4.1b in pairs: in the first zone, one pair is driven by the electromechanical relay and two pairs are controlled by the same solid-state relay. Instead, in the second zone, the EMR and the SSR handle one pair each. We denote the voltage drops across the heat resistors as $V_i^{(r)}(t) \in \mathbb{R}_{\geq 0}$ (in V), where $i \in \{1, 2\}$ denotes the zone while $r \in \{1, 2\}$ indicates the driving relay ($r = 1$ for the SSRs and $r = 2$ for the EMRs). Let $R_\mathrm{heat} \in \mathbb{R}_{>0}$ (in Ω) be the resistance of a pair of heat resistors and assume that it is the same for each pair. Then, due to Joule heating, considering the number of pairs of heating elements per zone:

$$q_1(t) = \frac{1}{R_\mathrm{heat}} \cdot \left[2 \cdot V_1^{(1)}(t)^2 + V_1^{(2)}(t)^2\right], \tag{4.6a}$$

$$q_2(t) = \frac{1}{R_\mathrm{heat}} \cdot \left[V_2^{(1)}(t)^2 + V_2^{(2)}(t)^2\right]. \tag{4.6b}$$

The voltages $V_i^{(r)}(t)$ in (4.6) are *PWM signals* with corresponding duty cycles $u_i^{(r)}(t) \in [0, 1]$. In particular, as pointed out in Sect. 4.1, the solid-state relays and the electromechanical relays operate at different periods: $\mathcal{T}_\mathrm{P}^{(1)} = 1$ s for the SSRs and $\mathcal{T}_\mathrm{P}^{(2)} = 30$ s for the EMRs. To ensure correct operation, the duty cycles must be held constant in each PWM time interval (e.g., via a zero-order holder), i.e.,

$$u_i^{(r)}(t) = u_i^{(r)}\left(k \cdot \mathcal{T}_\mathrm{P}^{(r)}\right), \quad \forall t : k \cdot \mathcal{T}_\mathrm{P}^{(r)} \leq t < (k+1) \cdot \mathcal{T}_\mathrm{P}^{(r)}, k \in \mathbb{N} \cup \{0\}. \tag{4.7}$$

Now, let $V_g(t) \in \mathbb{R}_{\geq 0}$ (in V) be the grid voltage. The voltage drop across each heat resistor pair amounts to

$$V_i^{(r)}(t) = \begin{cases} V_g(t) & \text{for } k \cdot \mathcal{T}_P^{(r)} \leq t < \left[k + u_i^{(r)}(t)\right] \cdot \mathcal{T}_P^{(r)} \\ 0 & \text{for } \left[k + u_i^{(r)}(t)\right] \cdot \mathcal{T}_P^{(r)} \leq t < [k+1] \cdot \mathcal{T}_P^{(r)} \end{cases}. \tag{4.8}$$

To avoid an uneven allocation of power among the heat resistors in each zone, the duty cycles $u_1^{(1)}(t)$, $u_1^{(2)}(t)$, and $u_2^{(1)}(t)$, $u_2^{(2)}(t)$, are forced to be (roughly) equal. To accomplish this, the temperature controller produces only two duty cycles (one *control action* per zone), denoted as

$$\boldsymbol{u}(t) = \left[u_1(t)\ u_2(t)\right]^\top \in [0, 1]^2, \tag{4.9}$$

every $\mathcal{T}_c \in \mathbb{R}_{>0}$ seconds (see Fig. 4.2), i.e., it computes $\boldsymbol{u}\left(k' \cdot \mathcal{T}_c\right)$ for every $k' \in \mathbb{N} \cup \{0\}$. Then, each $u_i^{(r)}(t)$, $i \in \{1, 2\}$, $r \in \{1, 2\}$, is updated every $\mathcal{T}_P^{(r)}$ seconds based on $u_i(t)$ in (4.9). Given that $\mathcal{T}_P^{(1)} \neq \mathcal{T}_P^{(2)}$, the sampling time $\mathcal{T}_c$ cannot be set arbitrarily. Furthermore, we need ways to derive $u_i^{(r)}\left(k \cdot \mathcal{T}_P^{(r)}\right)$ from the values of $u_i\left(k' \cdot \mathcal{T}_c\right)$ obtained by the controller for all $k, k' \in \mathbb{N} \cup \{0\}$. That is the purpose of the following *Operation Modes (*`OM`*s)*, which are at the core of the duty cycle allocation block in the scheme in Fig. 4.2:

- `OM1`: $\mathcal{T}_c = \mathcal{T}_P^{(1)}$, i.e., the controller updates the control actions at the end of every period of the PWM signals produced by the SSRs. In this case, the duty cycles related to the SSRs match the control actions, while those of the EMRs are computed by averaging the $u_i(t)$'s in the past $\gamma = \frac{\mathcal{T}_P^{(2)}}{\mathcal{T}_P^{(1)}} = 30$ periods of the SSRs (see Fig. 4.4a). Formally, $\forall i \in \{1, 2\}$, we have

$$u_i^{(1)}\left(k \cdot \mathcal{T}_P^{(1)}\right) = u_i\left(k' \cdot \mathcal{T}_c\right), \quad \forall k \in \mathbb{N} \cup \{0\}, k' = k, \tag{4.10a}$$

$$u_i^{(2)}\left(k \cdot \mathcal{T}_P^{(2)}\right) = \begin{cases} \frac{1}{\gamma} \cdot \sum_{k'=0}^{\gamma-1} u_i\left((k-1) \cdot \mathcal{T}_P^{(2)} + k' \cdot \mathcal{T}_c\right) & \forall k \in \mathbb{N} \\ 0 & \text{for } k = 0 \end{cases}. \tag{4.10b}$$

- `OM2`: $\mathcal{T}_c = m \cdot \mathcal{T}_P^{(2)}$, $m \in \mathbb{N}$, i.e., the controller computes the control actions at a time multiple of the period of the PWM signals produced by the EMRs. In this case, the duty cycles related to the SSRs and EMRs match the control actions and are held constant for one or more PWM periods (see Fig. 4.4b). Formally, for $k, k' \in \mathbb{N} \cup \{0\}$ and $\forall i \in \{1, 2\}$, we have

$$u_i^{(1)}\left(k \cdot \mathcal{T}_P^{(1)}\right) = u_i\left(k' \cdot \mathcal{T}_c\right), \quad \forall k : \gamma \cdot m \cdot k' \leq k < \gamma \cdot m \cdot (k'+1), \tag{4.11a}$$

$$u_i^{(2)}\left(k \cdot \mathcal{T}_P^{(2)}\right) = u_i\left(k' \cdot \mathcal{T}_c\right), \quad \forall k : m \cdot k' \leq k < m \cdot (k'+1). \tag{4.11b}$$

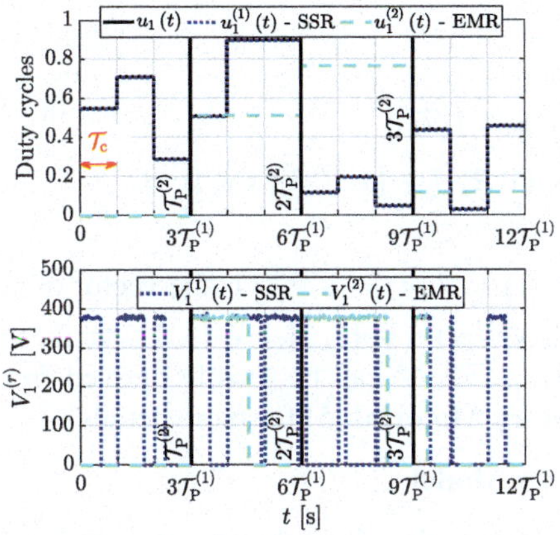

(a) OM1, $\mathcal{T}_c = \mathcal{T}_P^{(1)} = 1$ s.

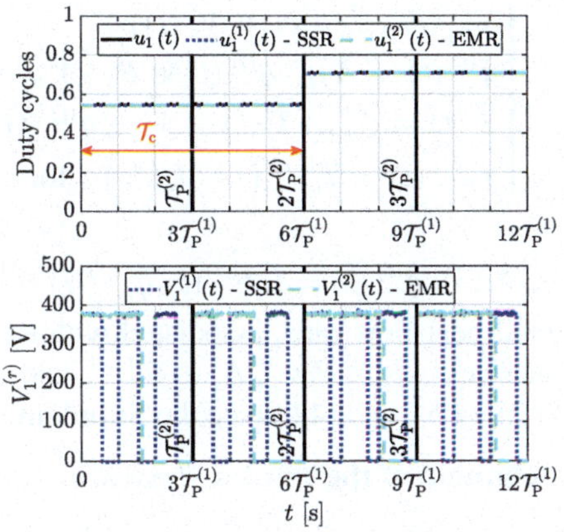

(b) OM2, $\mathcal{T}_c = 2 \cdot \mathcal{T}_P^{(2)} = 6$ s ($m = 2$).

Fig. 4.4 Example of PWM signals for the duty cycle $u_1(t)$ generated by the controller and with $\mathcal{T}_P^{(1)} = 1$ s, $\mathcal{T}_P^{(2)} = 3$ s, for the two different operation modes in (4.10) and (4.11). Typically, the grid voltage $V_g(t)$ is not constant, as highlighted in the figure

As a final remark, the choice of the operation mode and, consequently, $\mathcal{T}_c$ depend on the complexity of the temperature controller, as detailed in Sect. 4.3.

To go on with the derivation of the `Heat-Temperatures` model, let $\boldsymbol{V}(t) \in \mathbb{R}^4_{\geq 0}$ be the vector of voltages and $\boldsymbol{V}_{\text{sq}}(t) \in \mathbb{R}^4_{\geq 0}$ be the vector of squared voltages, i.e.,

$$\boldsymbol{V}(t) = \left[V_1^{(1)}(t)\ V_1^{(2)}(t)\ V_2^{(1)}(t)\ V_2^{(2)}(t)\right]^\top,$$
$$\boldsymbol{V}_{\text{sq}}(t) = \boldsymbol{V}(t) \odot \boldsymbol{V}(t).$$

For ease of discussion, we define a function $\boldsymbol{f} : [0, 1]^2 \times \mathbb{R}_{\geq 0} \to \mathbb{R}^4$ that describes the relationship between the control actions and the squared voltages:

$$\boldsymbol{V}_{\text{sq}}(t) = \boldsymbol{f}\left(\boldsymbol{u}(t), V_g(t)\right). \tag{4.12}$$

$\boldsymbol{f}\left(\boldsymbol{u}(t), V_g(t)\right)$ is obtained by composition of (4.8), taking into account the two previously described operation modes in (4.10) and (4.11), and the square of each $V_i^{(r)}(t)$. Finally, we can combine the system of differential equations in (4.2), the time lag introduced in (4.4), the filtered heat flow rates in (4.5), the relationship between the $q_i(t)$'s and the squared voltages $V_i^{(r)}(t)^2$ in (4.6), and (4.12), which describes how the voltage signals are shaped, to obtain the `Heat-Temperatures` model:

$$
\begin{cases}
\boldsymbol{V}_{\text{sq}}(t) = \boldsymbol{f}\left(\boldsymbol{u}(t), V_{\text{g}}(t)\right) \\
\dot{\boldsymbol{q}}^{(\text{f})}(t) = A_{ff} \cdot \boldsymbol{q}^{(\text{f})}(t) + B_f \cdot \boldsymbol{V}_{\text{sq}}(t) \\
\dot{\boldsymbol{T}}(t) = A_{TT} \cdot \boldsymbol{T}(t) + A_{Tf} \cdot \boldsymbol{q}^{(\text{f})}(t) + \boldsymbol{b}_T \cdot T_{\text{a}}(t) \\
y_{T_i}^{(l,s)}(t) = T_i^{(l,s)}\left(t - L_{T_i}^{(l,s)}\right) \quad \text{for } i \in \{1,2\}, l \in \{1,2,3\}, s \in \{\text{l}, \text{r}\}.
\end{cases}
\tag{4.13}
$$

In (4.13), $\boldsymbol{q}^{(\text{f})}(t) = \left[q_1^{(1,\text{f})}(t)\ q_1^{(2,\text{f})}(t)\ q_2^{(1,\text{f})}(t)\ q_2^{(2,\text{f})}(t)\right]^\top \in \mathbb{R}_{\geq 0}^4$ is the vector of filtered heat flow rates associated with each $V_i^{(r)}(t), i \in \{1,2\}, r \in \{1,2\}$, while the matrices $A_{ff} \in \mathbb{R}^{4\times 4}$, $B_f \in \mathbb{R}^{4\times 4}$, and $A_{Tf} \in \mathbb{R}^{12\times 4}$ can be readily derived from (4.2), (4.5), and (4.6) (see the **Supplementary Material** [5] for more details).

Definition of the `Packs-Temperatures` Model

In this section, we derive a model that describes the effect that the flow of bottle packs has on the air temperature inside the oven. Before being inserted into the cavity, the bottles are at the ambient temperature or lower, which is less than the air temperature within the heating zones. Therefore, some of the heat inside the cavity will be absorbed by both the beverages and the plastic film wrapped around them, effectively lowering the temperatures inside the oven. The speed at which the bottle packs arrive at the entrance of the cavity is dictated by the production rate of the industrial process which produces the beverages. In what follows, we denote as $v(t)$ the production rate at time t which can either be 0, 30, 60, or 90 ppm (packs per minute), i.e., $v(t) \in \{0, 30, 60, 90\}$. We can detect the presence (1) or absence (0) of bottle packs at the entrance of the oven through the infrared sensor in Fig. 4.1a. We denote as $d(t; v(t)) \in \{0, 1\}$ the reading of the infrared sensor at time t. We model $d(t; v(t))$ using a *Bernoulli distribution* [9] with a success rate that depends on the production rate. Intuitively, when $v(t)$ is high, it is more likely that there are bottles at the entrance of the oven cavity. Let $p_d(v(t)) \in [0, 1]$ be the production-rate-dependent success rate, then:

$$
d(t; v(t)) \sim \text{Bernoulli}\left(p_d(v(t))\right). \tag{4.14}
$$

Clearly, $p_d(v(t)) = 0$ when $v(t) = 0$, i.e., there are no bottle packs at the entrance of the oven cavity when no beverages are being produced.

Let $\delta T_i^{(l,s)}(t) \in \mathbb{R}$ (in °C), $i \in \{1,2\}, l \in \{1,2,3\}, s \in \{\text{l}, \text{r}\}$, denote the temperature drop at each spot in Fig. 4.1b due to the insertion of bottle packs inside the oven cavity. From experimental insights derived from the shrink tunnel under study, we model the relationship between $d(t; v(t))$ and each $\delta T_i^{(l,s)}(t)$ using a first-order low-pass filter with gain $\mu_{d_i}^{(l,s)}(v(t)) \in \mathbb{R}_{<0}$ (in °C) and time constant $\tau_{d_i}^{(l,s)}(v(t)) \in \mathbb{R}_{>0}$ (in s). Similar to (4.14), both parameters depend on the production rate. Formally:

$$
\delta \dot{T}_i^{(l,s)}(t) = -\frac{1}{\tau_{d_i}^{(l,s)}(v(t))} \cdot \delta T_i^{(l,s)}(t) + \frac{\mu_{d_i}^{(l,s)}(v(t))}{\tau_{d_i}^{(l,s)}(v(t))} \cdot d(t; v(t)). \tag{4.15}
$$

We group all the temperature drops $\delta T_i^{(l,s)}(t)$ inside a vector $\delta \boldsymbol{T}(t) \in \mathbb{R}^{12}$, ordered as in (4.3). Then, analogously to (4.4), we introduce a production-rate-dependent time lag $L_{d_i}^{(l,s)}(v(t)) \in \mathbb{R}_{\geq 0}$ (in s) and define the `Packs-Temperatures` model as follows:

$$\begin{cases} \delta \dot{\boldsymbol{T}}(t) = A_{dd}(v(t)) \cdot \delta \boldsymbol{T}(t) + \boldsymbol{b}_d(v(t)) \cdot d\,(t; v(t)) \\ y_{d_i}^{(l,s)}(t) = \delta T_i^{(l,s)}\left(t - L_{d_i}^{(l,s)}(v(t))\right) \quad \text{for } i \in \{1,2\}, l \in \{1,2,3\}, s \in \{\text{l}, \text{r}\}. \end{cases} \tag{4.16}$$

The matrix $A_{dd}(v(t)) \in \mathbb{R}^{12\times 12}$ and vector $\boldsymbol{b}_d(v(t)) \in \mathbb{R}^{12}$ can be readily derived from (4.15) and are reported in the **Supplementary Material** [5].

Complete Shrink Tunnel Model

The complete model for the shrink tunnel under study can be obtained by combining the `Heat-Temperatures` model in (4.13), the `Packs-Temperatures` model in (4.16), and (4.1), leading to

$$\begin{cases} \boldsymbol{V}_{\text{sq}}(t) = \boldsymbol{f}\left(\boldsymbol{u}(t), V_{\text{g}}(t)\right) \\ \dot{\boldsymbol{q}}^{(\text{f})}(t) = A_{ff} \cdot \boldsymbol{q}^{(\text{f})}(t) + B_f \cdot \boldsymbol{V}_{\text{sq}}(t) \\ \dot{\boldsymbol{T}}(t) = A_{TT} \cdot \boldsymbol{T}(t) + A_{Tf} \cdot \boldsymbol{q}^{(\text{f})}(t) + \boldsymbol{b}_T \cdot T_{\text{a}}(t) \\ \delta \dot{\boldsymbol{T}}(t) = A_{dd}(v(t)) \cdot \delta \boldsymbol{T}(t) + \boldsymbol{b}_d(v(t)) \cdot d\,(t; v(t)) \\ y_{T_i}^{(l,s)}(t) = T_i^{(l,s)}\left(t - L_{T_i}^{(l,s)}\right) \qquad \text{for } i \in \{1,2\}, l \in \{1,2,3\}, s \in \{\text{l}, \text{r}\} \\ y_{d_i}^{(l,s)}(t) = \delta T_i^{(l,s)}\left(t - L_{d_i}^{(l,s)}(v(t))\right) \\ y_i^{(l,s)}(t) = y_{T_i}^{(l,s)}(t) + y_{d_i}^{(l,s)}(t). \end{cases} \tag{4.17}$$

Overall, (4.17) represents a nonlinear state-space model with twenty-eight states, i.e., $\boldsymbol{x}(t) = \left[\boldsymbol{q}^{(\text{f})}(t)^\top\ \boldsymbol{T}(t)^\top\ \delta\boldsymbol{T}(t)^\top\right]^\top \in \mathbb{R}^{28}$, two inputs (the control actions), namely $\boldsymbol{u}(t) \in [0,1]^2$, four external disturbances, which are $V_{\text{g}}(t) \in \mathbb{R}_{\geq 0}$, $T_{\text{a}}(t) \in \mathbb{R}$, $v(t) \in \{0, 30, 60, 90\}$, and $d\,(t; v(t)) \in \{0, 1\}$, and twelve outputs $\boldsymbol{y}(t) = \left[y_1^{(3,\text{r})}(t) \cdots y_2^{(3,\text{l})}(t)\right]^\top \in \mathbb{R}^{12}$ (ordered as in (4.3)). All the parameters for the model in (4.17) are reported in the **Supplementary Material** [5].

4.2.2 Control Objectives and Constraints

The control objectives and the constraints on the control actions that ensure the correct and optimal operation of the shrink tunnel under study are:

- *Setpoint Tracking* (and *temperature uniformity*): When the oven is initially turned on, the air inside it is at the ambient temperature $\bar{T}_{\text{a}}$. Before inserting any bottle pack, the oven must reach an operating point that is between 150 °C and 170 °C depending on the type of beverages that will be inserted into it and the plastic film wrapped around the bottles. A constant setpoint signal $\text{SP}(t) = \overline{\text{SP}}, \forall t \in$

$\mathbb{R}_{\geq 0}$, $\bar{\text{SP}} \in \mathbb{R}$ (in °C), is given as a reference to the controller. The regulator must generate the duty cycles $u_1(t)$ and $u_2(t)$ to drive all the temperatures inside the cavity to be as close as possible to the setpoint. Clearly, there is a *controllability problem* since there are fewer control actions than outputs. Therefore, the controller cannot steer all the temperatures toward the setpoint. Typically, it is more important that the temperatures near the center of the cavity (i.e., $y_i^{(l,s)}(t)$, $i \in \{1, 2\}$, $l \in \{1, 2\}$, $s \in \{\text{l}, \text{r}\}$, in Fig. 4.1b) follow SP (t) more closely than those at the entrance and exit (i.e., $y_1^{(3,s)}(t)$, $y_2^{(3,s)}(t)$, $s \in \{\text{l}, \text{r}\}$, in Fig. 4.1b) because that is where the plastic shrinking process takes place. In any case, it is desirable to have temperature uniformity within the oven cavity.

- *Disturbance Rejection:* To meet quality specifications regarding the plastic wrapping surrounding the bottle packs coming out of the shrink tunnel, deviance from the equilibrium temperatures[1] must be kept at a minimum, especially when $v(t) \neq 0$. As a rule of thumb, when the temperatures deviate from their equilibrium values by 5 °C or more (mostly for what concerns the temperatures at the center of the oven), the resulting bottle packs will be discarded due to poor quality of the wrapping. Due to the high production rates (30, 60, or 90 packs per minute), not meeting the ±5 °C tolerance for just a few minutes may lead to substantial economic loss for the company owning the shrink tunnel.
- *Energy Saving:* Shrink tunnels can operate uninterruptedly for several hours or days before being turned off. Therefore, to minimize economic loss, the controller should also minimize the electrical energy consumption of the heat resistors.
- *Constraints on the Control Actions:* The duty cycles $u_1(t)$ and $u_2(t)$ computed by the controller must be such that

$$0 \leq u_i(t) \leq 1, \quad \forall t \in \mathbb{R}_{\geq 0}, i \in \{1, 2\}, \tag{4.18}$$

otherwise the voltage PWM signals in (4.8) are not defined correctly. Moreover, the heat resistors of the considered shrink tunnel are designed to operate at a nominal voltage of $V_{\text{nom}} = 380$ V, leading to a maximum nominal power of

$$P_{\max} = \frac{V_{\text{nom}}^2}{R_{\text{heat}}}.$$

In practice, the grid voltage $V_{\text{g}}(t)$ might differ from V_{nom} and may vary over time. Consequently, the controller must generate the duty cycles $u_1(t)$ and $u_2(t)$ in such a way that the average power supplied to the heat resistors within a period of PWM does not exceed $P_{\max}$. This amounts to imposing

$$\frac{1}{\mathcal{T}_{\text{P}}^{(r)}} \cdot \int_{k \cdot \mathcal{T}_{\text{P}}^{(r)}}^{(k+1) \cdot \mathcal{T}_{\text{P}}^{(r)}} \frac{V_i^{(r)}(t)^2}{R_{\text{heat}}} dt \leq P_{\max}, \quad \forall i \in \{1, 2\}, r \in \{1, 2\}, k \in \mathbb{N} \cup \{0\}. \tag{4.19}$$

[1] In general, the equilibrium values may differ from the setpoint due to the aforementioned controllability issue.

Notice that (4.19) constitutes an additional constraint on $u_1(t)$ and $u_2(t)$ since the voltages $V_i^{(r)}(t)$ depend on the duty cycles, as highlighted in (4.8). The integral in (4.19) depends on the future values assumed by $V_g(t)$ in the time interval $k \cdot \mathcal{T}_P^{(r)} \leq t < (k+1) \cdot \mathcal{T}_P^{(r)}$, which are not available at $t = k \cdot \mathcal{T}_P^{(r)}$. Let $u_i^{(r)}\left(k \cdot \mathcal{T}_P^{(r)}\right)$ be the duty cycles of the voltage PWM signals in that time interval (see (4.7)). Assuming that $V_g(t)$ stays roughly constant in each PWM period, we can make the following approximation:

$$\int_{k \cdot \mathcal{T}_P^{(r)}}^{(k+1) \cdot \mathcal{T}_P^{(r)}} V_i^{(r)}(t)^2 \, dt = \int_{k \cdot \mathcal{T}_P^{(r)}}^{\left[k + u_i^{(r)}\left(k \cdot \mathcal{T}_P^{(r)}\right)\right] \cdot \mathcal{T}_P^{(r)}} V_g(t)^2 \, dt$$
$$\approx V_g\left(k \cdot \mathcal{T}_P^{(r)}\right)^2 \cdot u_i^{(r)}\left(k \cdot \mathcal{T}_P^{(r)}\right) \cdot \mathcal{T}_P^{(r)}$$

and re-write the constraint in (4.19) accordingly.

4.3 Benchmark Design

We benchmark the temperature controller for the shrink tunnel under study by simulating a *typical working day for a bottle packs manufacturing process*. In particular, we consider a constant setpoint $\bar{\text{SP}} = 160\,°\text{C}$ and a production schedule with several rate changes and breaks; see Fig. 4.5. The grid voltage fluctuates between 380 V and 410 V. Initially, the air inside the oven is at the ambient temperature $\bar{T}_a = 15\,°\text{C}$.

The user must design a temperature controller that satisfies the constraints in (4.18) *and* (4.19), *and guarantees the stability of the system*. Notice that the controller in Fig. 4.2 updates the control actions $u_1(t)$ and $u_2(t)$ every $\mathcal{T}_c$ seconds based on the signals $\mathbf{y}(t)$, $d(t; v(t))$, and $V_g(t)$, which are sampled at a sampling time $\mathcal{T}_c$. Consequently, *the user must first select an operation mode* (see Sect. 4.2.1) *and then tune* $\mathcal{T}_c$ *accordingly*: either $\mathcal{T}_c = \mathcal{T}_P^{(1)} = 1$ s for `OM1` in (4.10), or $\mathcal{T}_c = m \cdot \mathcal{T}_P^{(2)} = m \cdot$

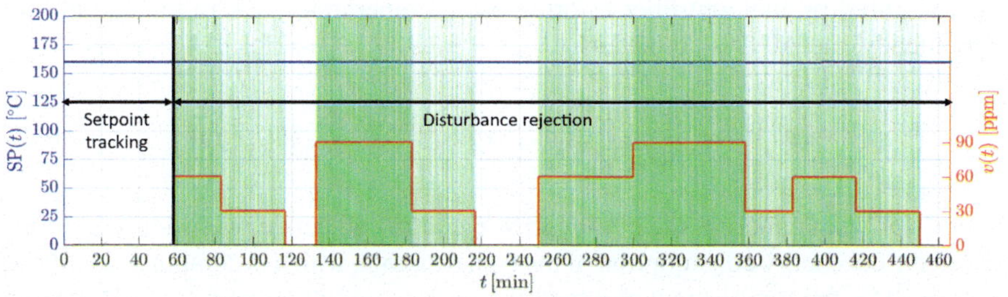

Fig. 4.5 Signals of interest for the benchmark: setpoint SP (t) and production rate $v(t)$. The green vertical lines denote when the infrared sensor at the entrance of the oven detects a bottle pack (i.e., when $d(t; v(t)) = 1$)

30 s, $m \in \mathbb{N}$, for OM2 in (4.11). Note that the computational burden of the regulator should be taken into account when choosing $\mathcal{T}_c$. Computationally light controllers, e.g., PIDs [1], can be run at $\mathcal{T}_c = \mathcal{T}_P^{(1)}$, while more time-consuming controllers, such as Model Predictive Controllers (MPCs) [8], can be executed at $\mathcal{T}_c = m \cdot \mathcal{T}_P^{(2)}$ with m tuned accordingly.

The user-defined controller must steer all the temperatures measured by the twelve thermocouples in Fig. 4.1b toward the equilibrium before the bottle packs production starts. Consistently with the Heat-Temperatures model in (4.13), we denote the temperature equilibrium values due to the heat generated by the heat resistors and the ambient temperature as $\bar{y}_{T_i}^{(l,s)} \in \mathbb{R}$, $i \in \{1, 2\}$, $l \in \{1, 2, 3\}$, $s \in \{\text{l}, \text{r}\}$. Given that the setpoint is constant, to assess the *setpoint tracking performance* of the controller we use common step response indicators [4]:

- The *settling time* of each $y_{T_i}^{(l,s)}$:

$$t_{\text{S}_i}^{(l,s)} = \frac{1}{60} \cdot \min_{\tau \in \mathbb{R}_{\geq 0}} \tau \text{ s.t. } \left\{ \left| y_{T_i}^{(l,s)}(t) - \bar{y}_{T_i}^{(l,s)} \right| \leq 0.02 \cdot \left| \bar{y}_{T_i}^{(l,s)} \right|, \forall t \geq \tau \right\} \quad [\text{min}]. \tag{4.20}$$

- The *absolute steady-state errors*:

$$e_{\text{SS}_i}^{(l,s)} = \left| \bar{\text{SP}} - \bar{y}_{T_i}^{(l,s)} \right| \quad [^\circ\text{C}]. \tag{4.21}$$

In practice, we want to minimize both $t_{\text{S}_i}^{(l,s)}$ in (4.20) and $e_{\text{SS}_i}^{(l,s)}$ in (4.21) to achieve fast response times and a uniform temperature distribution within the oven cavity. For what concerns the *disturbance rejection performance* of the controller, we define an indicator according to the corresponding control objective reported in Sect. 4.2.2. Let

$$t_A > \max_{i \in \{1,2\}, l\{1,2,3\}, n \in \{\text{l},\text{r}\}} t_{\text{S}_i}^{(l,n)}$$

be the time at which the first bottle pack is inserted into the oven cavity and $t_B > t_A$ be the last. We want to minimize the *Area Under the Curve (AUC)* performance indicator, which accounts for both the magnitude of the temperature drop and the time required by the controller to bring each temperature back to the equilibrium. Specifically, the AUC is defined as

$$\text{AUC}_i^{(l,s)} = \frac{1}{60} \cdot \int_{t_A}^{t_B} \left| y_i^{(l,s)}(t) - \bar{y}_{T_i}^{(l,s)} \right| dt \quad [^\circ\text{C} \cdot \text{min}]. \tag{4.22}$$

Lastly, for what concerns the *energy saving* control objective, we want to minimize the electrical energy consumption of the heat resistors, which is computed as follows. Let $t = 0$ be the time instant at which the oven is turned on and $t_{\text{stop}} \in \mathbb{R}_{>0}$ be the time at which the shrink tunnel is shut down (i.e., $u_1(t) = u_2(t) = 0, \forall t > t_{\text{stop}}$). Then, the energy consumption is given by

$$E_V = \frac{1}{3600} \cdot \sum_{i=1}^{2} \sum_{r=1}^{2} \int_0^{t_{\text{stop}}} \frac{V_i^{(r)}(t)^2}{R_{\text{heat}}} dt \quad [\text{W} \cdot \text{h}]. \tag{4.23}$$

4.4 Accessing the Benchmark

This chapter's benchmark was developed in MATLAB® and Simulink®. In particular, in the **Supplementary Material** [5], we provide two Simulink® models and two MATLAB® scripts that allow the user to simulate the shrink tunnel under study:

1. `complete_model_ol.slx` is the Simulink® model representing the shrink tunnel when operated in open loop. In the MATLAB® script `simulate_OL.m`, the user defines the duty cycles $u_1(t)$ and $u_2(t)$, the grid voltage $V_g(t)$, and the production rate $v(t)$. Then, they can run `simulate_OL.m` to simulate the behavior of the system under study and analyze the resulting outputs $\boldsymbol{y}(t)$.
2. `complete_model_cl.slx` is the Simulink® model representing the shrink tunnel when operated in closed loop. Therein, a subsystem reference block to accomodate the user-defined controller has already been created with all the necessary inputs and outputs. The user must design a Simulink® subsystem that represents their controller for the system under study. When defining the regulator, the user must account for the control specifications described in Sects. 4.2.2 and 4.3. The MATLAB® script `run_benchmark.m` plugs in the user's controller into the `complete_model_cl.slx` model and runs the benchmark, deriving the performance indicators presented in Sect. 4.3. The signals $\text{SP}(t)$, $V_g(t)$, and $v(t)$ are already defined in `run_benchmark.m` according to Sect. 4.3.

In any case, the user must also select the operation mode and the sampling time T_c as described in Sect. 4.3. The Simulink® models have been coded following the block diagram in Fig. 4.2. Relevant documentation is provided in the **Supplementary Material** [5]. Therein, the user can also find a simple PID control strategy as an example.

4.5 Discussion for Future Comparison

The shrink tunnel model proposed in this chapter is novel and therefore (currently) there are no scientific papers covering possible control architectures that tackle the control objectives and constraints presented in Sect. 4.2.2. However, there are similar applications in the control literature, which we cover here. The works presented in this section address the temperature control of different thermal systems, such as buildings and environmental test chambers, focusing on setpoint tracking, temperature uniformity, disturbance rejection, and energy saving. We review three works that differ in the employed control architecture and models.

In [3], the authors consider a temperature test chamber with eight distributed heaters, each with its own temperature sensor, resulting in a multivariable system with eight inputs and eight outputs. For control purposes, the relationship between each input and each output is approximated by a *black-box First-Order Lag Plus time Delay (FOLPD) transfer function*. Then, the authors design a control architecture composed of *two sets of PID controllers* with the objectives of setpoint tracking and temperature uniformity. The resulting control actions are the sum of the contributions produced by the aforementioned regulators. The first set of PIDs aims to make the average temperature track the setpoint signal based on the average error. Instead, the second set of PIDs improves the uniformity of the temperatures inside the test chamber by considering only the difference between the tracking error and the average error.

In [7], the authors employ *Successive linear MPC (SMPC), Linear Time-Varying MPC (LTV-MPC), and Nonlinear MPC (NMPC)* for the temperature control of a radiant-floor building. The system is modeled by the composition of several *gray-box submodels*, such as the building, heat pump, and hot water tank models; each submodel is derived from physics, using first principles. Ultimately, the assembled model is a continuous-time nonlinear time-varying system in state-space form. The control objectives are setpoint tracking and energy saving. In particular, all the considered MPC regulators rely on a cost function that takes into account the electricity cost due to the heat pump power consumption and the discrepancies between the building air temperature and the setpoint. Moreover, all MPC strategies can maintain the comfort conditions in the building closely.

In [12], the authors design an advanced control algorithm for the cooling system of a large business and commercial center. This complex system is governed by phenomena that are difficult to model with physics principles, such as the demand of the users, the heat transport phenomena in large and complex pipe networks, and the behavior of the cooling elements installed by third parties. These numerous aspects of uncertainty motivate the usage of a black-box machine learning approach, such as *Long Short-Term Memory (LSTM) neural networks*, to model the dynamical behavior of the system under study. The control objectives are multiple and sometimes conflicting, such as (i) keeping the collector water within a certain temperature range, (ii) avoiding the chiller's activation for short periods (to prevent an economic loss), and (iii) reducing the number of active chillers to decrease energy consumption. The control system is composed of an *Extended Kalman Filter (EKF) and a SMPC*. The control system's performance shows that the cooling system's main temperatures are maintained within prescribed ranges while limiting energy consumption.

References

1. Astrom KJ, Hagglund T (1995) PID controllers: theory, design, and tuning. The instrumentation, systems, and automation society

2. Sure Controls (2014) Sure controls' relay selector guide. https://www.surecontrols.com/wp-content/uploads/2014/07/HA136716-CII-Issue-2c-continental-1.pdf. Accessed 06 Jul 2023
3. He W, Guanhua X, Shen R (2014) Control of temperature uniformity in the temperature chamber with centrifugal acceleration. J Process Control 24(12):1–6
4. Ogata K (2010) Modern control engineering, vol 5. Prentice hall Upper Saddle River, NJ
5. Previtali D, Pitturelli L, Previdi F, Ferramosca A (2024) Supplementary material. https://github.com/Lenyor/Temperature_control_of_a_shrink_tunnel_with_multiple_heating_zones
6. Ramirez-Laboreo E, Sagues C, Llorente S (2016) Dynamic heat and mass transfer model of an electric oven for energy analysis. Appl Therm Eng 93:683–691
7. Rastegarpour S, Gros S, Ferrarini L (2020) MPC approaches for modulating air-to-water heat pumps in radiant-floor buildings. Control Eng Pract 95:104209
8. Rawlings JB, Mayne DQ, Diehl M (2017) Model predictive control: theory, computation, and design, vol 2. Nob Hill Publishing Madison, WI
9. Ross SM (2020) Introduction to probability and statistics for engineers and scientists. Academic press
10. Serale G, Fiorentini M, Capozzoli A, Bernardini D, Bemporad A (2018) Model predictive control (MPC) for enhancing building and HVAC system energy efficiency: problem formulation, applications and opportunities. Energies 11(3):631
11. Sidebotham G (2015) Heat transfer modeling. Springer
12. Terzi E, Bonetti T, Saccani D, Farina M, Fagiano L, Scattolini R (2020) Learning-based predictive control of the cooling system of a large business centre. Control Eng Pract 97:104348
13. Xin X, Liu Y (2014) Control design and analysis for underactuated robotic systems. Springer

Chapter 5
A Model for Gas Humidification in a Fuel-Cell Assembly

Mateo Arcila-Osorio, Hernán Alvarez, and Carlos Ocampo-Martinez

Abstract This chapter proposes a novel model for the gas humidification process in a fuel-cell assembly as a benchmark for estimation and control-oriented tasks. The text delves into the benchmark's origins, its importance, and the challenges it poses to control engineers. The problem description offers a detailed overview of the assembly designed to provide gases to a fuel-cell with controlled humidity and temperature. The deduced mathematical model is described, emphasizing the use of a Phenomenological Based Semi-physical Modeling (PBSM) approach, which is an intuitive and simple method for deducing models. Key assumptions and simplifications are outlined for clarity. The chapter also details the experimental setup and the performance metrics used for evaluating control systems, providing a comprehensive view of the benchmark's design and associated limitations. The access to the benchmark's resources is easy through a web repository, containing both MATLAB® and Python scripts to solve the whole model. Potential users are guided on how to access and run the benchmark. Future research directions and existing approaches to solving this problem are discussed, identifying gaps in the current state of the art. Finally, the references cited in the chapter are listed to enrich the reader's understanding.

5.1 Introduction

The control of humidity and temperature of a gas stream in a process line is a challenge due to the strongly coupled behavior of these two variables and their nonlinear interaction. Besides, for streams of gases or vapors the effect of the stream pressure on

M. Arcila-Osorio (✉) · C. Ocampo-Martinez
Automatic Control Department, Universitat Politècnica de Catalunya, Barcelona, Spain
e-mail: mateo.arcila@upc.edu

C. Ocampo-Martinez
e-mail: carlos.ocampo@upc.edu

H. Alvarez
Universidad Nacional de Colombia, Medellín, Colombia
e-mail: hdalvare@unal.edu.co

J. M. Maestre and C. Ocampo-Martinez (eds.), *Control Systems Benchmarks*, Advances in Industrial Control, https://doi.org/10.1007/978-3-031-76312-0_5

humidity is very strong and must be considered while looking for a right execution of modeling and control tasks. When pressure changes, the gas capacity to retain vapor is modified. For instance, an increase of pressure can bring the gas stream closer to its saturation value suddenly. Regarding temperature changes, when the temperature of the gas–vapor mixture increases, the gas capability for retaining humidity also increases. Despite being controlled, temperature experiences disturbances from the surroundings, affecting humidity and pressure. The intensely thermodynamic relationship among these variables is evident. Obviously, assuming a constant pressure for all the lines conducting the gas or vapor reduces the model complexity. However, for process lines of low diameter and assembled with several fittings as valves, reductions, contractions, etc., this assumption is very restrictive, leading to imprecise models. As is well-known, if the model does not have enough precision, a model-based control strategy using this model will offer a poor performance.

The current benchmark appears after dealing with the difficult task of controlling a pilot Proton Exchange Membrane (PEM) fuel-cell assembly used for energy production. Looking for providing an accurate enough tool for humidification and temperature control, this benchmark presents a model for the process of conditioning the supply gases for the fuel-cell. The relative humidity and temperature of the gases supplied to a fuel-cell must be controlled with enough precision to guarantee the proper cell operation and its usable life. Humidity values outside the narrow range of acceptable values can lead to cell flooding or damage in the materials forming the cell. Excessive or insufficient temperature relative to the set value results in a significant drop in cell performance, leading the device to not satisfy the current electric load demands. Therefore, an adequate control of humidity and temperature of the gases supplied to the fuel-cell is a mandatory task when the efficient operation of this equipment is the goal.

Fuel-cells hold great promise for automotive applications due to their high efficiency and low emissions. However, challenges persist, including high costs primarily attributed to materials like platinum, durability concerns under harsh conditions, and the need for a comprehensive hydrogen infrastructure. Safety considerations regarding hydrogen storage and transportation, as well as cold start performance, also present hurdles. Further, improving efficiency and performance remains crucial for enhancing fuel-cell vehicles' competitiveness with conventional options. Addressing these challenges requires ongoing research and development efforts across various disciplines to unlock the full potential of fuel-cell technology in the automotive sector.

The main control problem to be solved using this benchmark is to profit the high interaction between humidity, temperature, and pressure in gas and vapor process streams. Consequently, any control system regulating these variables must consider such strong interaction in order to execute a satisfying process control. The thermodynamic relationship between these variables is well-established, but the challenge for the controller arises from the difficulties in measuring the values of these variables at various points along the process line. In that sense, the use of the benchmark presented here will enable the implementation of model-based strategies for virtual sensors or controllers that take advantage of the strong coupling between humidity, temperature, and pressure, rather than concealing it. Certainly, the estimation of values of variables along the conducting line will provide enough information to propose an advanced control strategy. However, the selected control strategy should fulfill the requirement of a nonlinear and highly coupled process.

To close this section, it is worth mentioning the work of Melo et al. [1], which presents a comprehensive discussion about the uses of benchmarks for control tasks, particularly oriented to process monitoring and fault diagnosis techniques. Some general characteristics of benchmarks are addressed, such as the richness of information, ease of use, availability of real data, challenges that must be considered, and the potential for scientific advances. The authors give a critical review that is highly valuable for any reader interested in using a benchmark for research on control strategies or testing these strategies to improve their application in industrial processes.

5.2 Problem Description

The process considered in this chapter is the assembly of pieces of equipment used to provide air, with regulated humidity and temperature, to either the cathode or the anode of a fuel-cell. The assembly, illustrated in Fig. 5.1, consists of a compressor, a gas filter, two electric heaters (for both liquid water and air), and a commercial membrane-based humidifier (Cellkraft P series 10). All these elements are connected through pipes and fittings of known diameter, configuration, and materials (Teflon hose and stainless steel tube). The most complex device in this assembly is the humidifier. This equipment is a sealed unit consisting of a bank of Nafion membrane tubes into a stainless steel cylindrical shield, arranged like a tubes-and-shield heat exchanger. Liquid water fills the shield space until a certain level, which is reached by using a local regulator. The air to be humidified flows through the tubes. The Nafion is Poly-Tetra-Fluor-Ethylene (PTFE) known commercially as Teflon, a material highly resistant to chemical reactions and permeable to water. Nafion allows a regulated mass transfer from liquid water to gas according to the membrane temperature [2]. In order to perform mass transfer manipulation, an electric-band heater for liquid water is externally installed at the bottom of the humidifier unit shield, covering a third of its cylindrical shield surface.

The mathematical model is derived using a combination of both white-box and black-box models through a Phenomenological Based Semi-physical Modeling (PBSM) approach. In brief, this methodology involves establishing a clear modeling objective, defining Process Systems (PSs) of interest, and constructing the model basic structure using first principles, i.e., material, momentum, and energy balances applied to the selected PSs. Subsequently, appropriate constitutive and assessment equations, forming the model extended structure, are employed to evaluate both structural and functional parameters. A computer program is then developed to simultaneously solve the model equations. Finally, the acquired data is compared with experimental data under specific operating conditions, constituting the model validation. This approach is fully illustrated in [3] and it has been used in several works due to its capabilities, including the production of models with generalization capacity, modularity, and interpretability of parameters.

The phenomena occurring in the gas humidification process are three, momentum exchange causing the flow of gases, heat exchange from heaters to flowing gases, and mass transfer of water to air in the humidifier. All of these phenomena are well-known and extensively documented in the literature [4]. The model objective is to represent changes in pressure, temperature, and humidity of the air flowing through the assembly just before the PEM fuel-cell, when main process inputs (temperature, pressure, and humidity of the gas supply to the entire assembly) suffer bounded variations. Within the scope of this benchmark, a lumped-parameter model is formulated at the macroscopic level of detail.

The declared PSs in accordance to the mass, heat, and momentum transfers that take place in the gas humidifier process are shown in Fig. 5.2. In this figure, thin arrows indicate mass flows and thick arrows represent energy flows. Electric heaters are not considered as PSs due to their simple function as power supplies. A brief description for each PS follows. The indicated points are marked in Figs. 5.1 and 5.2.

- PS_I is the air contained into the stainless-steel tube and fittings assembly, from point 0 to point 1.
- PS_{II} represents the air contained into the plastic hose, from point 1 to point 2.
- PS_{III} indicates the air contained into the stainless-steel tube and fittings assembly, extending from point 2 to 3.
- PS_{IV} is the air contained into the plastic hose, from point 3 to point 4.
- PS_V indicates the air contained into the Nafion tubes of the humidifier, from point 4 to point 5.

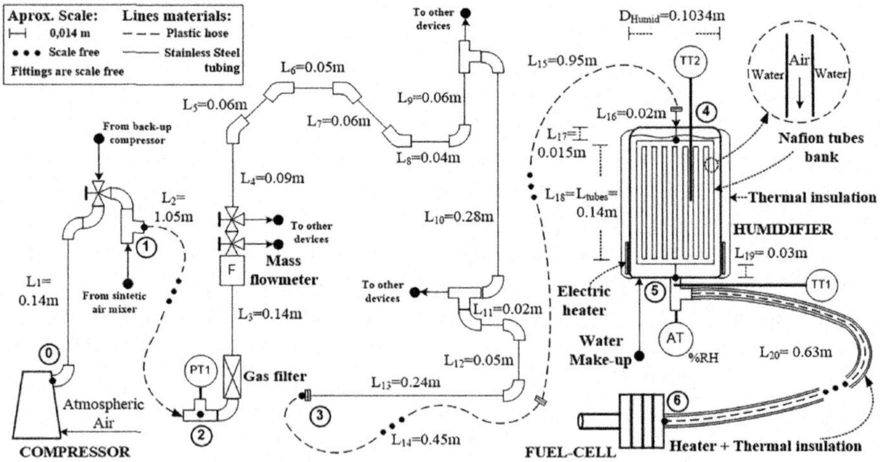

Fig. 5.1 General flow diagram for the air-humidification process

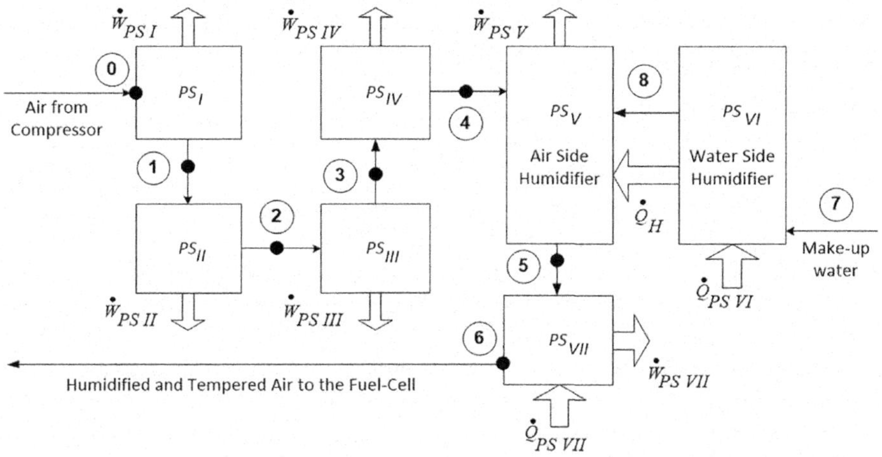

Fig. 5.2 Block diagram of the proposed process systems (PSs)

- PS_{VI} represents the liquid water contained within the humidifier shield.
- PS_{VII} is the air contained into the heated plastic hose, extending from point 5 to point 6. This PS corresponds to the system that must be controlled.

The detailed specification given to each section of air conduction should be highlighted, looking for a precise evaluation of pressure losses and air conditions at different points of the assembly. In the sequel, the conservation principle is applied to each one of these PSs, providing valuable information to fulfill the model objective. In that sense, the list of assumptions used to derive the model structure, all being valid simplifications for this process, is outlined as follows:

A1. There is not mechanical work (shaft work) at any part of the process.
A2. The heat exchanged with the surroundings is directly subtracted from the electric heater power during the parameter identification.
A3. Liquid water stream has enough pressure to flow through the humidifier, therefore no mechanical energy balances are required in this stream.
A4. The humidified air is an incompressible fluid and an ideal gas due to the low-pressure drops experienced during its pass through the process.
A5. Mechanical equilibrium of fluid streams is immediate due to the small diameter and the short length of the tube line, but the pressure losses are not null while the gas is passing through the tubes and fittings.
A6. The Nafion membrane forming the humidifier does not have appreciable heat exchange effects. Therefore, its role on mass transfer is only due to net temperature gradient.

The subsequent balance equations constitute the model basic structure. The deduction and further simplifications to determine these final expressions are available in the supplementary material (see Sect. 5.4). Additionally, it is possible to access the extended structure of the model, including used constant values and all parameter constitutive and assessment equations.

- Process Systems I to IV:

$$\dot{m}_{i+1} = \dot{m}_i, \tag{5.1a}$$

$$P_{i+1} = P_i - \rho_{a_i}\, g\, (z_i - z_{i+1}) - \rho_{a_i}\, h_{f_{i\to i+1}}, \tag{5.1b}$$

being $\dot{m}_i$ [kg/s] the mass flow rate of gas mixture and $i = 0, ..., 3$ an index for the stream numbers into circles, in accordance with the block diagram in Fig. 5.2. Besides, P_i [Pa] is the pressure of the stream evaluated at point i, ρ_{a_i} [kg/m^3] is the density of the dry air and water vapor mixture, g [m/s^2] is the gravity acceleration constant, z_i [m] corresponds to the height of current point with respect to a reference point in the assembly, and $h_{f_{i\to i+1}}$ [m^2/s^2] represents the friction losses due to the gas mixture flow between points i and $i+1$.

- Process System V:

$$\frac{dM_{PS_V}}{dt} = \dot{m}_4 - \dot{m}_5 + \dot{m}_8, \tag{5.2a}$$

$$\frac{dw_{wv,5}}{dt} = \frac{1}{M_{PS_V}}\Big[w_{wv,4}\, \dot{m}_4 - w_{wv,5}\, \dot{m}_5 + \dot{m}_8 - w_{wv,5}\, \dot{M}_{PS_V}\Big], \tag{5.2b}$$

$$\frac{dT_5}{dt} = \frac{1}{\widehat{C}_{V_{a,5}}\, M_{PS_V}}\Big[-\dot{m}_8\, \widehat{C}_{P_{wv,8}}\, (T_5 - T_8) + \dot{Q}_{T-A}\Big], \tag{5.2c}$$

$$P_5 = P_4 - \Delta P_{4\to 5}, \tag{5.2d}$$

where M_{PS_V} [kg] is the total amount of substance contained in PS_V, $\dot{m}_4$ and $\dot{m}_5$ are the inlet and outlet mass flow rates of gas mixture, respectively, and $\dot{m}_8$ is the mass transferred from the water side to the air side within the humidifier. Similarly, $w_{wv,i}$ represents the water vapor mass fraction in stream i, $\dot{M}_{PS_V} = \dot{m}_4 - \dot{m}_5 + \dot{m}_8$ at any particular time, T_i [K] is the temperature of the stream i, $\widehat{C}_{V_{a,5}}$ [kJ/kg K] is the specific heat capacity at constant volume of the humid air at stream 5 conditions, and $\widehat{C}_{P_{wv,8}}$ [kJ/kg K] corresponds to the specific heat capacity at constant pressure of the water vapor at stream 8 conditions. Lastly, $\dot{Q}_{T-A}$ [kJ/s] expresses the heat transferred from hot Nafion tubes into the humidifier shield to the flowing humid air inside the Nafion tubes, and $\Delta P_{4\to 5}$ [Pa] accounts for all pressure changes of humid air flowing through the humidifier tubes.

- Process System VI:

$$\dot{m}_7 = \dot{m}_8, \tag{5.3a}$$

$$\frac{dT_8}{dt} = \beta\Big[-\dot{m}_8\, \lambda_{Vap} - \dot{m}_8\, \widehat{C}_{P_{wL}}\, (T_8 - T_7) + \dot{Q}_{H1-W} - \dot{Q}_{T-A} - \dot{Q}_{Lost}\Big], \tag{5.3b}$$

with $\beta = \left(\widehat{C}_{P_{wL}}\ M_{PS_{VI}}\right)^{-1}$ and with $\widehat{C}_{P_{wL}}$ [kJ/kg K] being the specific heat capacity at constant pressure of liquid water. Moreover, $M_{PS_{VI}}$ [kg] is the amount of substance contained in PS_{VI}, λ_{Vap} [kJ/kg] is the vaporization heat of water, $\dot{Q}_{H1-W}$ [kJ/s] is the heat transferred from the electric heater of the humidifier to the liquid water, and $\dot{Q}_{Lost}$ [kJ/s] indicates the heat lost from the humidifier to the surrounding environment.

- Process System VII:

$$\dot{m}_6 = \dot{m}_5, \tag{5.4a}$$

$$\frac{dT_6}{dt} = \frac{1}{\widehat{C}_{V_{a,6}}\ M_{PS_{VII}}}\left[\dot{m}_6\ \widehat{C}_{P_{a,6}}\ (T_5 - T_6) + \dot{Q}_{H2-A}\right], \tag{5.4b}$$

$$P_6 = P_5 - \rho_{a,5}\ h_{f_{5\to 6}}, \tag{5.4c}$$

where T_6 [K] is the outlet temperature of the humid air from the assembly (the fuel-cell inlet), $M_{PS_{VII}}$ [kg] represents the humid air mass within the tube, and $\widehat{C}_{V_{a,6}}$ [kJ/kg K] and $\widehat{C}_{P_{a,6}}$ [kJ/kg K] are the specific heat capacities of humid air at constant volume and at constant pressure in stream 6, respectively. Finally, $\dot{Q}_{H2-W}$ [kJ/s] expresses the heat transferred from the external heater installed around this assembly section.

After developing all the model equations, the computational model was directly programmed and solved in both MATLAB® and Python using the Euler method with a 0.01 s fixed step. This step is crucial for a successful simulation; using this short time allows to reconstruct the strong nonlinear effects caused by fast changes on flows and temperatures at the low mass flow rates of this process. Furthermore, in order to analyze the behavior of the proposed dynamic model, available data reported in [5] have been used, particularly the test with a power of 123 W applied to the humidifier heater and a flow of 5 NL/min (NL for liters of gas at Normal conditions) of gas mixture feeding the assembly. The results from the comparisons, including relative humidity, temperature, and pressure at the humidifier outlet, validated the ability of the dynamic model to reproduce the actual evolution of the key variables measured in this particular case study.

In the context of the correct design and implementation of a control strategy, the benchmark users must not lose sight of the scope of their proposed control system. Therefore, this section concludes by emphasizing important aspects and providing some guidance in this regard. Hence, as mentioned earlier, to avoid poor cell performance and mitigate potential damage to the fuel-cell membrane under both dry and flooded operating conditions, it is imperative to ensure that the supplied gases maintain precise humidity and temperature conditions. Consequently, the system to be controlled consists of the humid air that flows in the section of pipe between the humidifier outlet and the fuel-cell inlet. The controlled variables are the relative humidity and the temperature of the air entering the fuel-cell, while the humidifier acts as the final control element to regulate these variables. Namely, the manipulated variable is the heater power supply. The purpose of the control system is to minimize the variability in the controlled variables in response to possible disturbances. This ensures stable operation despite external factors affecting the process.

Finally, it is important to mention the main operational constraints of the process, which are related to the operating range of the humidifier. The Cellkraft P-10 model allows an inlet gas flow from 0 to 10 slpm (standard liter per minute), a temperature within the interval from 20 to 300°C, a gas pressure between 0 and 20 bar, and a heater power up to 1200 W. Under these conditions, the relative humidity of the outlet gas can be set throughout its range from 0 to 100% [6].

5.3 Benchmark Design

The design of the proposed benchmark is not complicated, comprising an assembly for a gas conditioning stage. The experimental setup for this benchmark was described in Sect. 5.2. This is not an extremely simple setup, but it can be accessible from available pieces of equipment, conduction lines, and fittings. There are no relevant limitations for implementing this assembly at laboratory scale. However, the following operational considerations are recommended:

- The level of water into the humidifier should be guaranteed. The current benchmark considers this level as a constant.
- The temperature of the last section of the line should be regulated by using a control loop.
- The gas inlet pressure should be guaranteed from the compressor or using a pressure regulator when the gas supply is a pressurized container.
- In case of replacing the water inside the humidifier with another liquid to enhance heat exchange properties, some parameters and constants of the proposed model should be suitably tuned before considering the use of both the model and the setup.
- Certain fuel-cell operations might influence the operation of the humidifier, requiring particular responses from the main device. This fact should be taken into account when updating the humidifier model for the scenarios under consideration.

This benchmark is similar to others proposed in the control literature [7–9], but in this case the two controlled variables have a strong interaction without mass or energy recycling. Despite the absence of recycling in this assembly, the strong and nonlinear relation between humidity, pressure, and temperature, in addition to the mass transfer phenomena included in the humidifier, poses a challenge to the modeling and control practitioner. Although this complexity suggests a confusing model, the current benchmark employs a simple model that is easy to understand in terms of its operation and the involved phenomenology. This fact allows the model to be used in an undergraduate control course without requiring a deep conceptualization of the phenomena. It is enough to understand the thermodynamic relationship between humidity, pressure, and temperature in a gas or vapor. In this way, the benchmark can initially be used for modeling tasks, but opening the door to model-based control tasks for advanced students or control practitioners. The use of proper automatic control in this assembly ensures the suitable performance of the fuel-cell, preserving the functionality of the different pieces of equipment constituting the assembly.

The suggested controller performance metrics are the typical ones used in control systems. Any implemented control strategy must ensure process stability, meaning that bounded perturbations yield bounded outputs. However, it is well-known that stability alone does not guarantee adequate performance. This is why various other well-established metrics exist. Benchmark users are encouraged to explore diverse performance metrics that they find relevant to their studies. For example, but not limited to, the Integral Time Absolute Error (ITAE) can assess how effectively the control strategy maintains the desired setpoints for temperature and relative humidity, while the rise time, overshoot, and setting time are suitable to evaluate how quickly the system responds to disturbances and whether it overshoots or oscillates before settling into the desired conditions. The steady-state error assesses how closely the system approaches the desired setpoints once it reaches steady-state conditions, and robustness evaluates its response capacity to disturbances, model uncertainties, or variations in operating conditions. Besides, the time out of a band around the setpoint can be considered. The recommended band around the setpoint is 10% for temperature and 5% for humidity and pressure. Certainly, narrowing these bands will pose a greater challenge to the tested control system. It should be highlighted that the control method for the heater power supply can be replaced with any user-defined strategy directly in the benchmark computational model.

Regarding modeling tasks, the performance metric is the Mean Absolute Error (MAE) of the model, concerning the previously mentioned data [5] used for model validation. The objective is to reduce the current prediction errors, which are 0.25% for humidity, 0.1°C for temperature, and 110 Pa for pressure. These errors were calculated using the values obtained after 3000 s of operation time, following the initial system stabilization and its approach to steady-state conditions. While one proposal to accomplish this objective is to implement any model improvement from the phenomenological-based approach [10], readers are encouraged to explore other approaches based on emergent strategies exploiting the availability of raw data (e.g., learning-based techniques).

The recommended scenarios to test the benchmark as a controlled process or as a process being modeled are (*i*) an increase or decrease of 20% in the inlet gas pressure, (*ii*) 15% increase or decrease in the gas humidity at the assembly inlet, and (*iii*) 5% increase or decrease in the gas inlet temperature, expressed in K. Besides, a complementary scenario implies the assumption of changes in the mass transfer process within the humidifier. The suggestion is to reduce the value of the mass transfer coefficient by 15% due to the possible Nafion membrane malfunction (fouling or porosity reduction). These values are specified as reasonable ranges that guarantee the stability of the open-loop dynamic response. However, with the implementation of an appropriate control strategy, it becomes possible to test disturbances of greater magnitude while maintaining stability in the system response.

The operating conditions of a PEM fuel-cell can vary significantly and depend primarily on the manufacturer and the specific application for which they were designed. Typically, proper operation is found in a temperature range from 60 to 80°C and between 40 and 100% of gas relative humidity. Within the context of this benchmark, users can play with these operating conditions and switch between different temperature and relative humidity setpoints to evaluate their control strategies. Any further changes to the process variables should be made while considering the normal range of values for these variables. For all previous scenarios, the variables to be measured in the gas are the pressure, temperature, and humidity at the fuel-cell inlet. In addition, the benchmark provides intermediate values of the variables, which can be used for diagnosing malfunctions of the gas conditioning process.

5.4 Accessing the Benchmark

The supplementary material for this benchmark can be found in a web repository.[1] These additional resources and files provide more details regarding the formulation of the basic structure of the model, the extended structure of the model, the computational model developed in both MATLAB® and Python, and some guidelines for effective implementation by the user. The source codes are freely available and can be downloaded and run to generate process data.

The repository is publicly accessible on GitHub, allowing anyone with Internet access to retrieve its contents. MATLAB® users must possess an active license for software usage, along with a compatible version to execute live scripts. Python users have the flexibility to run the provided code in any free Python Integrated Development Environment (IDE), including Spyder, Jupyter, and Google Colab. It is particularly advisable to select the latter two, given that the source codes were developed in the format of a Python notebook. These notebooks serve as interactive documents capable of incorporating live code, equations, visualizations, and narrative text.

The model presented in this chapter serves as a data generation system, specifically providing time series data that allows users to observe the dynamic behavior of the variables and parameters relevant to the process. Benchmark users can analyze the values of the controlled variables (temperature T_6, relative humidity RH_6, and pressure P_6 at the fuel-cell inlet), the manipulated variables (the heater power supply $\dot{Q}_{H1\text{-}W}$ and $\dot{Q}_{H2\text{-}A}$), and the main disturbances (temperature T_{Amb}, relative humidity RH_{Amb}, and pressure P_0 of the gas supply to the entire assembly) under different circumstances. This data is available in vector form and can be visualized through plots for further clarity. For instance, Fig. 5.3 illustrates the relative humidity and temperature responses at the fuel-cell inlet following a 15% positive step change over the humidity at the assembly inlet, after 6000 s of operation time. Additionally, users can access values for other parameters, such as pressures at the different points marked in Fig. 5.2 (see Sect. 5.2), gas density within each process system, mass transfer flux

[1] https://github.com/arcmateo/GasHumidifierBenchmark.git.

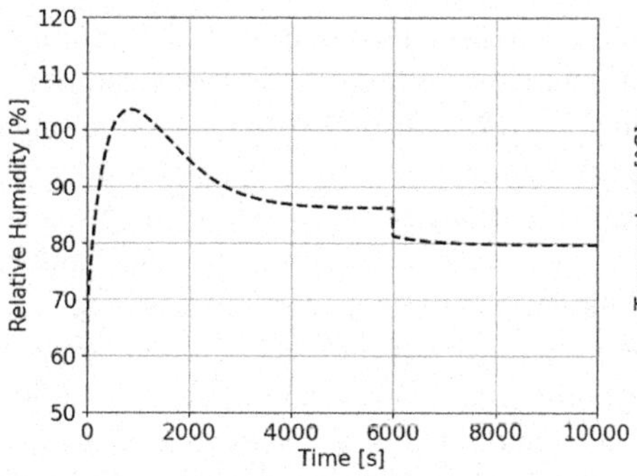

(a) Humidity response at fuel-cell inlet.

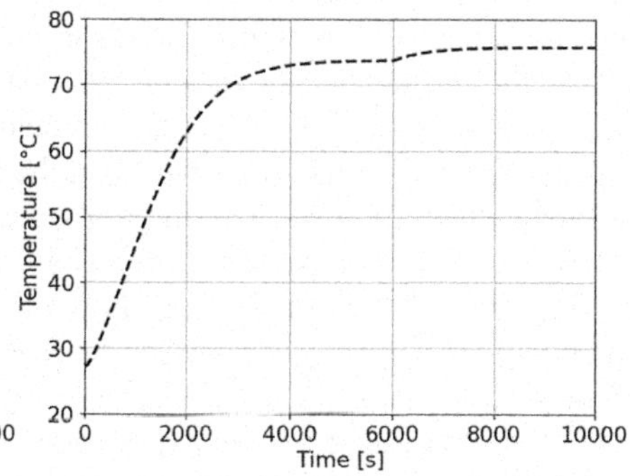

(b) Temperature response at fuel-cell inlet.

Fig. 5.3 Example of dynamic response of key variables (humidity and temperature) after a disturbance (15% inlet humidity positive step change at $t = 6000$ s)

in the humidifier, and so forth. By modifying operating conditions, introducing step changes to input variables, and programming disturbances, users can collect multiple datasets. Subsequently, during the implementation of the control strategy, identical datasets can be acquired from the closed-loop system. As a result, the controller's efficacy can be rigorously assessed employing the performance metrics.

5.5 Discussion for Future Comparison

Some approaches to the humidification process for gases to be used in a fuel-cell exist, but none of them is available as a benchmark. Chen and Peng [11] provide a detailed model for the membrane humidifier operation. The model includes the effect of temperature, pressure, and relative humidity of the air at the humidifier inlet. Steady-state and dynamic simulations are performed to optimize the humidifier design. A simple proportional (P) controller for humidity is tested, but nothing is discussed about the strong interaction of the mentioned variables in the assembly out of the humidifier. Liu et al. [12] present a humidification control-oriented model that focuses on the mass transfer process, but without delving into the detailed effects of pressure and temperature on the gas humidity. A feedforward control proposal is tested for the humidification process under high current demand, improving the performance of the humidifier without affecting the regulation of the oxygen excess ratio. However, this work does not provide a comprehensive discussion on the potential limitations and uncertainties associated with the assumptions made in the modeling and control development. Zhang et al. [13] operated with a process model for the air supply system of a fuel-cell assembly constructed by a commercial software complemented with Simulink® of MATLAB®. The attention of this work is on the compressor, its performance, and gas characteristics analysis. Nothing is mentioned about the temperature and pressure effect over gas humidity. Ma et al. [14] proposed a model for a fuel-cell test platform humidifier. The paper focuses on the humidifier, its operation, and modeling details, including the computational domain, boundary conditions, and simulation models used. The influence of operating conditions such as inlet air velocity and water temperature on the humidification performance is addressed. Although the introduction of this work briefly mentions the temperature and pressure effects on gas humidity, these effects are not incorporated into the model. Wang et al. [15] studied the effect of gas flow on the humidifier performance in a fuel-cell assembly. The proposed model includes the effects of dry and wet bulb temperature of the air, as well as the gas pressure in the humidifier. However, the effects of tubing and fitting forming the assembly are not considered. Finally, the authors indicate that further research will be conducted to analyze how temperature, pressure, and flow channel affect the performance of the humidifier.

Clearly, the benchmark presented in this chapter holds significant value as it is designed to address some of these challenging gaps. In addition, the model is outlined in detail with its deduction and using a simple and intuitive methodology.

Acknowledgements The authors would like to thank the project PID2023-148840OB-I00 SEAMLESS funded by MCIN/AEI/10.13039/501100011033/FEDER, UE and TED2021-129927B-I00 (MASHED) funded by MCIN/AEI/10.13039/501100011033 and by the "European Union NextGenerationEU/PRTR" for supporting this research.

References

1. Melo A, Câmara MM, Clavijo N, Carlos Pinto J (2022) Open benchmarks for assessment of process monitoring and fault diagnosis techniques: a review and critical analysis. Comput Chem Eng 165:107964
2. Kang S, Min K, Yu S (2012) Dynamic modeling of a proton exchange membrane fuel cell system with a shell-and-tube gas-to-gas membrane humidifier. Int J Hydrogen Energy 37(7):5866–5875; XII international symposium on polymer electrolytes: new materials for application in proton exchange membrane fuel cells
3. Lema-Perez L, Garcia-Tirado J, Builes-Montaño C, Alvarez H (2019) Phenomenological-based model of human stomach and its role in glucose metabolism. J Theor Biol 460:88–100
4. Bird RB, Stewart WE, Lightfoot EN (2007) Transport phenomena. Wiley, New York
5. Solsona M, Kunusch C, Ocampo-Martinez C (2017) Control-oriented model of a membrane humidifier for fuel cell applications. Energy Conv Manag 137:121–129
6. Cellkraft. Flow-through humidifiers. https://cellkraft.se/humidifiers/flow-through-humidifiers/
7. Alvarez H, Londoño C, di Sciascio F, Carelli R (2001) ph neutralization process as a benchmark for testing nonlinear controllers. Ind Eng Chem Res 40(11):2467–2473
8. Box GEP, Reinsel GC, Jenkins GM, Ljung GM (2015) Time series analysis: forecasting and control, 5th edn
9. Downs JJ, Vogel EF (1993) A plant-wide industrial process control problem. Comput Chem Eng 17(3):245–255
10. Lema-Perez L, Muñoz-Tamayo R, Garcia-Tirado J, Alvarez H (2019) On parameter interpretability of phenomenological-based semiphysical models in biology. Inf Med Unlocked 15:100158
11. Chen D, Peng H (2004) A thermodynamic model of membrane humidifiers for PEM fuel cell humidification control. J Dyn Syst Meas Control 127(3):424–432

12. Liu Z, Chen J, Chen S, Huang L, Shao Z (2017) Modeling and control of cathode air humidity for pem fuel cell systems. IFAC-PapersOnLine 50(1):4751–4756. 20th IFAC World Congress
13. Zhang Y, Bao P, Wan Y, Xu S (2017) Modeling and analysis of air supply system of polymer electrolyte membrane fuel cell system. Energy Procedia 142:1053–1058; Proceedings of the 9th International Conference on Applied Energy
14. Ma T, Wang K, Zhou Q, Lin W, Cong M, Jia W (2019) Numerical study on humidification performance of fuel cell test platform humidifier. Energies 12:3839
15. Wang X, Li H, Yin R, Li C (2029) Design of humidifier test system for proton exchange membrane fuel cell. J Phys Conf Ser 012074(09):2021

Chapter 6
A Benchmark for Plant-Wide Optimization and Control of Activated Sludge Wastewater Treatment Plants

Silvana Revollar, Mario Francisco, Montse Meneses, Pastora Vega, and Ramón Vilanova

6.1 Introduction

The Benchmark Simulation Models (BSM1, BSM1_LT, BSM2) have been designed as a reference for simulation-based comparison of plant-wide operation and control strategies in a municipal wastewater treatment plant (WWTP) [1, 2]. These models have been designed by IWA Task Group on Benchmarking and Control Strategies for WWTPs.

The WWTPs are public services oriented to mitigate the environmental impact of wastewater discharges on water bodies. The operation of these facilities involves the use of chemicals, consumption of energy, and production of sludge that must be stabilized to be finally disposed. The classical configuration of a WWTP includes three process lines: water line, sludge line, and gas line. They comprise different types of biological treatments where complex and nonlinear biological processes take place, with different process units that are interconnected by mass and energy flows. Their operation objectives focus on fulfilling regulations applied to treated

S. Revollar (✉) · M. Francisco · P. Vega
Department of Computer Science and Automation, Universidad de Salamanca, Salamanca, Spain
e-mail: srevolla@usal.es

M. Francisco
e-mail: mfs@usal.es

P. Vega
e-mail: pvega@usal.es

M. Meneses · R. Vilanova
Department of Telecommunications and Systems Engineering, Universitat Autònoma de Barcelona, Bellaterra, Spain
e-mail: montse.meneses@uab.cat

R. Vilanova
e-mail: ramon.vilanova@uab.cat

J. M. Maestre and C. Ocampo-Martinez (eds.), *Control Systems Benchmarks*, Advances in Industrial Control, https://doi.org/10.1007/978-3-031-76312-0_6

water discharges minimizing energy consumption and emissions to water, air, and soil. Moreover, the operation must consider the variations on influent characteristics and climate variability that affect the global performance of the process. Such characteristics have motivated the use of the WWTP as a benchmark to evaluate the possibilities of different control strategies. The successful strategies could be exploited by the wastewater treatment sector that requires innovative technological developments to improve their efficiency, or could be applied to other processes with similar characteristics.

Particularly, the complexity of the activated sludge process (ASP) has always motivated a great research effort to develop suitable controllers and optimize its operation, revealing the need of a standard platform for simulations including a plant layout, mathematical models, performance indices, and influent profiles, for comparing results. The first ideas for a WWTP benchmark started 30 years ago at the Instrumentation, Control and Automation Conference in Canada. The work was initiated within the framework of the European COST Action 682 "Integrated Wastewater Management", and it continued under COST 624, giving as a result a first scientific and technical report describing the benchmark [1, 3]. In this period, the BSM1 and its extension BSM1_LT for long-term evaluation were developed and released. The plant-wide benchmark BSM2 development was also initiated in this time that enlarges the scope adding pre-treatment (primary clarifying) and a sludge line with anaerobic digestion and other units for sludge conditioning [1, 4].

After the end of the COST Action, an IWA benchmarking Task Group was created for 3 years, but in the end taking more time to finish the proposed work. At present days, there is an active group of researchers that continue to improve the benchmark adding more refined models and other aspects to provide more realistic simulations.

The target audience for this benchmark is wide. People both from industry and academia may benefit of the platform. In academia, the benchmark simulation models can be used for objective comparison of different control strategies, to train students on process control, or even to demonstrate the use of control strategies to a potential customer. The use of only parts of the benchmark or individual units could also be useful for some users [1]. In addition to process control, economic and other environmental issues are also included in the benchmark by means of standardized quantifying indexes. The benchmark can also be used as a tool for performing a Life Cycle Analysis (LCA) of the plant.

The challenges of this benchmark are linked to the fact that it is a very complex process, and it is not easy to find suitable control strategies. In addition to the inherent nonlinearity of the microbiological processes, the existence of recycling flows cause high interaction among units, requiring the use of a multivariable control strategy. The typical approach in current plants is the use of PI controllers for dissolved oxygen control in the bioreactors or ammonium-based supervisory control for the plants with available sensors. Only a small percentage of plants use advanced controllers. On the other hand, the simulation itself is cumbersome because it is a stiff system with algebraic loops, and the choice of the numerical method to integrate the equations is crucial to have good results in a reasonable computational time.

6.2 Problem Description

This section focuses on BSM1 and BSM2 models as simulation platforms that include the simulation code, the definition of influent pattern, and the performance evaluation criteria. The BSM1 is designed to evaluate control performance of a carbon/nitrogen removal activated sludge process (ASP) of a WWTP in a short operation period (7 d). BSM2 platform represents water and sludge line of a typical WWTP, as an extension of BSM1 concept for long-term monitoring and performance evaluation of plant-wide control strategies (1 year period) [1, 5].

6.2.1 Plant Description: Benchmark Layout and Mathematical Models

In a typical WWTP, the process units are grouped in water line, sludge line, and gas line. The water line includes a primary settler for removing suspended solids, and the biological treatment that is normally the ASP. Nitrogen removal ASP is considered in BSM models, and pollutant removal occurs in a train of bioreactors with anoxic and aerobic zones through nitrification and denitrification processes. An internal recycle transports the nitrates produced in the aerobic zone to the anoxic one. Clean water is separated from the sludge in a secondary clarifier (or settler). Most of the sludge is recycled to the first reactor to maintain the adequate proportion between wastewater, and microorganisms (Food to Mass ratio, F/M). A small amount of sludge is withdrawn from the ASP to keep a suitable sludge age. This sludge in excess is mixed with primary settler underflow to be treated in the sludge line, where anaerobic digestion (AD) process is used to stabilize sludge converting most of the organic matter into a mixture of methane (CH_4), carbon dioxide (CO_2), and hydrogen (H_2), which reduces the amount of sludge for disposal. The mixture is biogas that is conditioned and stored in the gas line to produce energy for the plant. The WWTP performance is strongly affected by the frequent and large disturbances in the influent flow and load that depends on human activities, and meteorological and climatic conditions. In BSM benchmarks, the description of the influent characteristics is relevant for an appropriated simulation of WWTP behavior.

BSM1 Plant Layout: BSM1 plant consists of five biological reactors in series and a secondary settler (Fig. 6.1). The first two reactors are anoxic and the rest are aerobic. The internal recycle (Q_a) returns mixed liquor from the last reactor to the first one. The secondary settler separates the clean effluent (Q_e) from sludge. Settler underflow (Q_u) is divided into external recycle (Q_r) that returns sludge to the first reactor, and purge flow (Q_w) that is withdraw from the system. The influent data allows 14 days of simulation with sampling time of 15 min, for an evaluation period of one week.

BSM2 Plant Layout: Water line includes a primary settler and ASP system represented by an adapted BSM1 plant. Sludge line consists of a thickener for the

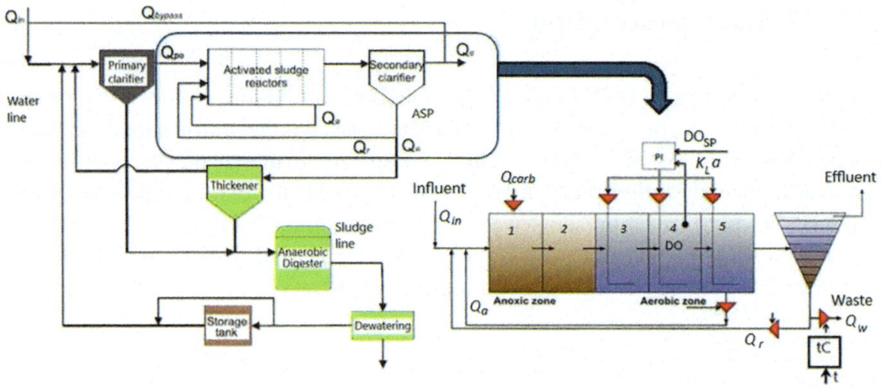

Fig. 6.1 Layout of a WWTP represented by benchmark BSM2 including ASP of BSM1 (Adapted from a combination of Figs. 1 and 2 in [4], released under a Creative Commons Attribution 4.0 International License (CC BY 4.0); https://wwtmodels.pubpub.org/pub/ytwymnp3)

Table 6.1 Volume of the BSM1 and BSM2 units

Process unit	BSM1 (m^3)	BSM2 (m^3)
Anoxic tank	1000	1500
Aerobic tank	1333	3000
Total ASP volume	5999	12000
Secondary settler	6000	6000
Primary settler	–	900
Anaerobic digester	–	3400
Storage tank	–	160

secondary sludge (Q_w), an anaerobic digester fed by a mixture of primary sludge and thickened sludge, a dewatering unit, and a storage tank that collects rejected water from the thickening and dewatering units which is recycled ahead of the primary settler (Fig. 6.1). For long-term evaluation with variable temperature, BSM2 uses a model generated influent profile that allows 609 days of simulation with a sampling time of 15 min. The evaluation period is one year (starting July 1st) in a plant located in the Northern Hemisphere. The recovering of energy from biogas is considered in the evaluation of operational costs. Process dimensions and operational parameters of BSM1 and BSM2 are summarized in Tables 6.1 and 6.2. The IWA Activated Sludge Model ASM1 [6] is typically used in BSM1 and BSM2 for ASP reactors, but in BSM1 constant temperature (15 °C) is considered and in BSM2 kinetic parameters and oxygen transfer depend on temperature [5]. General simplifications in BSMs are tanks perfectly mixed, and aeration intensity described by the oxygen transfer coefficient K_La, even in the aeration systems used for oxygen supply to aerobic reactors, where the airflow rate and the air bubble formation and diffusion determine

Table 6.2 Influent characteristics and operational parameters of BSMs plants

Parameter	BSM1	BSM2
Average influent flow rate (Q_{in})	18466 m^3/d	20648 m^3/d
Max influent flow (Q_{inmax})	60000 m^3/d	85841 m^3/d
Min influent flow (Q_{inmin})	10000 m^3/d	5146 m^3/d
Temperature (T)	15 °C	Average: 15 °C Span: [10–20] °C
Total ASP volume	5999 m^3	12000 m^3
Internal recycle flow (Q_a)	55338 m^3/d	61944 m^3/d
External recycle flow (Q_r)	18466 m^3/d	20648 m^3/d
Purge flow (Q_w)	385 m^3/d	300 m^3/d
Hydraulic retention time (HRT)	14.4 h	22 h
Sludge age	9 d	17.3 d

Table 6.3 Activated Sludge Model No. 1 (ASM1) variables

Definition	Notation	Definition	Notation
Soluble inert organic matter	S_I	Readily biodegradable substrate	S_S
Particulate inert organic matter	X_I	Slowly biodegradable substrate	X_S
Active heterotrophic biomass	$X_{B,H}$	Active autotrophic biomass	$X_{B,A}$
Particulate products from biomass decay	X_P	Oxygen	S_O
Nitrate and nitrite nitrogen	S_{NO}	NH_4^++NH_3 nitrogen (ammonium)	S_{NH}
Soluble biodegradable organic nitrogen	S_{ND}	Particulate biodegradable organic nitrogen	X_{ND}
Alkalinity	S_{ALK}		

aeration phenomena. Moreover, hydraulic delays have been added to ASP to prevent algebraic loops as a result of wastewater recycling [5].

The ASM1 represents the nitrification/denitrification mechanisms for nitrogen and organic matter removal considering 13 state variables (Table 6.3) and 8 biological processes (Table 6.4). In the nitrification process, the microbial biomass converts ammonium on nitrates in aerobic conditions. In the denitrification process, nitrates are reduced to nitrogen gas (N_2) in anoxic conditions. In the ASM1 model, the substrate is composed by organic and nitrogenous matter. Organic matter can be non-biodegradable (inert) and biodegradable that can also be soluble and particulate material, such as readily biodegradable substrate (soluble) and slowly biodegradable substrate (particulate). Similar classification applies to nitrogenous matter. The model distinguishes heterotrophic and autotrophic biomass and their growth is represented using Monod kinetics. The dependencies between the biological processes described in ASM1, and the state variables are presented in Table 6.4. As an example of the mathematical representation of biological processes, the anoxic growth of heterotrophs ρ_2, that depends on $X_{B,H}$, S_S, S_O, S_{NO} (see Table 6.3), is presented:

Table 6.4 ASM1 biological processes rates

Biological process	Process rate
Aerobic growth of heterotrophs	$\rho_1 = f(X_{B,H}, S_S, S_O)$
Anoxic growth of heterotrophs	$\rho_2 = f(X_{B,H}, S_S, S_O, S_{NO})$
Aerobic growth of autotrophs	$\rho_3 = f(X_{B,A}, S_O, S_{NH})$
Decay of heterotrophs	$\rho_4 = f(X_{B,H})$
Decay of autotrophs	$\rho_5 = f(X_{B,A})$
Ammonification of S_{ND}	$\rho_6 = f(X_{B,H}, S_{ND})$
Hydrolysis of entrapped organics	$\rho_7 = f(X_{B,H}, X_S, S_O, S_{NO})$
Hydrolysis of entrapped organic nitrogen	$\rho_8 = f(X_{B,H}, X_S, S_O, S_{NO}, X_{ND})$

$$\rho_2 = \mu_H \left(\frac{S_S}{K_S + S_S}\right)\left(\frac{K_{O,H}}{K_{O,H} + S_O}\right)\left(\frac{S_{NO}}{K_{NO} + S_{NO}}\right)\eta_g X_{B,H}. \tag{6.1}$$

The terms $S_S/(K_S + S_S)$ and $S_{NO}/(K_{NO} + S_{NO})$ represent the influence of S_S, S_{NO} as substrate in the process kinetics. Parameters K_S and K_{NO} are the semi-saturation constants that is the concentration for which the reaction rate is half of the maximum rate. The term $K_{OH}/(K_{OH} + S_O)$ is a control function which expresses the inhibitory effect of a certain substrate, in this case dissolved oxygen (DO, S_O) on the reaction [6]. The equations for mass balancing for each component are

$$\frac{dZ_1}{dt} = \frac{1}{V_1}(Q_a \cdot Z_a + Q_r \cdot Z_r + Q_0 \cdot Z_{po} + r_{z,1} \cdot V_1 - Q_1 \cdot Z_1) \quad \text{Reactor 1} \tag{6.2}$$

$$\frac{dZ_k}{dt} = \frac{1}{V_k}(Q_{k-1} \cdot Z_{k-1} + r_{z,k} \cdot V_k - Q_k \cdot Z_k) \quad \text{Reactors 2 to 5} \tag{6.3}$$

where k indicates reactor number, a internal recycle, r external recycle, and 0 the ASP input. Z is the concentration of any compound in the process, V_k is the volume in reactor k, Q_0 is the input flow rate, Q_1 is the combination of Q_a, Q_r, and Q_0, and Q_k is equal to Q_{k-1} (Fig. 6.1). The reaction rate (r_z) combines the processes involved in the consumption or growing of Z. No bulking and foaming processes are included, neither non-desirable reactions promoted by filamentous microorganisms that occur at DO concentration below 1.5 g O_2/m^3. For the special case of oxygen, an additional term is added to the mass balance equations, representing the oxygen transfer from liquid to gas phase, $K_La \cdot (So_{sat} - S_o)$, where So_{sat} is the saturation oxygen concentration and K_La is the oxygen transfer coefficient. A detailed description of the ASM1 model implementation in BSMs can be found in [3, 4]. BSM implementations including activated sludge models ASM2d and ASM3 are also available, and the details are in [7].

The secondary clarifier is a 10-layer sedimentation unit with different solid concentration. The feed layer is the 6th one, starting from the bottom layer [3, 4]. Takacs model is used to represent its behavior considering that settling properties of the sludge are not temperature-dependent [6].

In the sludge line, the anaerobic digester uses the IWA anaerobic digestion model No. 1 (ADM1) to describe the biochemical processes as hydrolysis, acidogenesis, acetogenesis, and methanogenesis, as well as physicochemical reactions as ion association/dissociation and gas–liquid transfer [8]. Liquid phase requires 26 equations for mass balances and 6 equations for the ion states, while gas phase requires 3 equations for CO_2, H_2, and CH_4. Some modifications have been donc to ADM1 to avoid numerical problems and an interface between states of ASM1 and ADM1 is necessary [1, 9]. The operation temperature is 35 °C, and a micro gas turbine is considered to produce the heat required by the digester and electricity [10]. The primary clarifier is described by a perfectly mixed tank. The thickener and dewatering units are modeled as a separation point with no volume. The storage tank is modeled as a nonreactive first-order system, with some conditions for extreme cases. The input and output streams from the different units can be observed in Fig. 6.1.

Regarding the influent model, the BSM1 provides three data files representing influent profiles with a constant influent temperature of 15 °C: dry weather, rain weather, and storm weather. The BSM2 influent profile (used also in BSM1_LT) is generated using a model that incorporates household and industrial pollutants. Influent model represents seasonal variations of temperature and precipitations, periodic variations during working days, weekends, and holidays [1, 11]. Sinusoidal functions are used to describe daily temperature variations with an amplitude of 1.75 °C, and seasonal variations between 20 °C and 10 °C, with an average of 15 °C. A detailed description of dynamic influent profile generation can be found in [11].

Sensors and actuators are represented using simplified mathematical models. The flow measurements are considered ideal. For other measures, six sensor classes are provided to be selected by users [1]. Most control handles are considered ideal, except for K_La that is modeled with a response time of 4 min, and a second-order delay function. Limits of the BSMs sensors and actuators are reported in [1].

6.2.2 Control Problem Description

The control problem focuses on the regulation of ASP variables that affect the effluent characteristics. From a plant-wide perspective, disturbance propagation and the interactions between different units of water and sludge line affect process efficiency, particularly in terms of energy usage. A default control strategy is provided in BSM1 and BSM2 to be used as a test case for comparisons [1, 2]; the control system parameters are shown in Table 6.5. The goal of the BSM1 control strategy is to maintain the S_{NO} concentration in the 2nd anoxic tank at 1 g N/m^3 manipulating the internal recycle (Q_a), and the DO concentration (S_O) in the 5th reactor at 2 g O_2/m^3 manipulating K_La_5, as shown in Fig. 6.2. This goal is attained using two PI control loops, and a gas-sensitive (class B_0) sensor for S_{NO} monitoring and an ion-sensitive sensor (class A) for S_O monitoring, both with measurement noise [1]. The BSM2 control

Table 6.5 Control system parameters of BSMs plants

BSM control system	Parameters
BSM1 Nitrate control	Controller gain: K = 10000 m^3/d/(g N/m^3)
PI anti-windup	Integral time constant: Ti = 0.025 d
	Anti-windup time constant: Tt = 0.015 d
BSM1, BSM2	Controller gain: K = 25 d^{-1}/(g -COD/m^3)
Oxygen control	Integral time constant: Ti = 0.002 d
PI anti-windup	Anti-windup time constant: Tt = 0.001 d
	Offset constant: 120 d^{-1}
BSM2	$0 \leq t < 182$ d $Q_w = 300\,m^3/d$
Sludge age control	182 d $\leq t < 364$ d $Q_w = 450\,m^3/d$
Timer control	364 d $\leq t < 546$ d $Q_w = 300\,m^3/d$
	546 d $\leq t < 608$ d $Q_w = 450\,m^3/d$

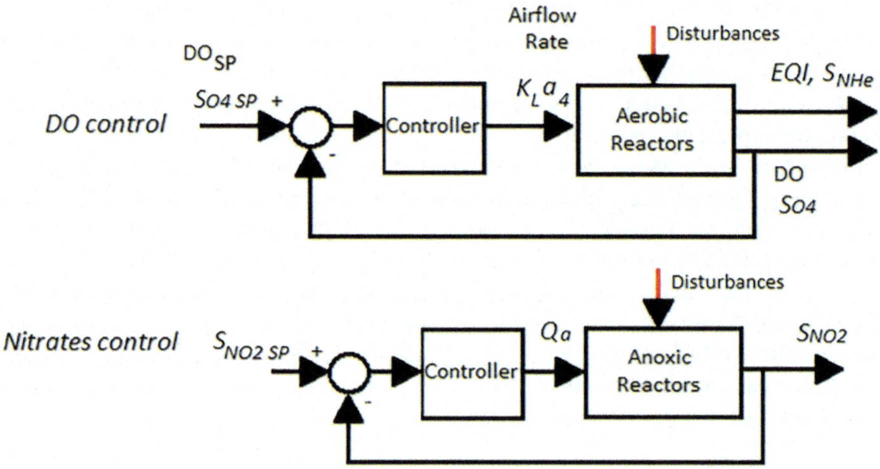

Fig. 6.2 BSM default control strategies. DO is S_{O4} in BSM2

strategy (Fig. 6.1) consists of the regulation of DO (S_O) in the 4th reactor at 2 g O_2/m^3 manipulating K_La_3, K_La_4, and K_La_5 (K_La_5 is set to half of K_La_3 and K_La_4 values) [1, 2]. Timer control of purge flow (Q_w) is implemented to adjust the sludge age to seasonal effects of temperature that affect biomass activity.

Other ASP control handles in BSMs are the sludge recycle flow (Q_r) and the external carbon source that is fed to the first anoxic reactor at a constant flow rate of 2 m^3/d and COD concentration of 400000 g/m^3 [1]. Moreover, mixing and aeration systems are represented for the five reactors which makes possible to change the volumes for anoxic and aerobic zones; the aeration intensity and the external carbon dosage can be modified individually for each tank, and the influent can be distributed fractionally to each of the tanks (step feed), as well as the Q_r and Q_a

flows [1]. In BSM2, additional control handles as the flow rates between units are also available [12].

Two alternative control strategies are mentioned in [1, 2]. A first one named CL1 is a modification of BSM2 default strategy, controlling the S_{O4} setpoint at 2 g O_2/m^3, by manipulating K_La_3 and K_La_4, while another loop regulates S_{O5} to 1 g O_2/m^3 by manipulating K_La_5. The other control strategy named CL2 adds a hierarchical control to CL1. A PI controller is applied to control S_{NH5} in the fifth reactor at a setpoint of 1.5 g N/m^3 by manipulating S_{O5} setpoint, and the external carbon flow rate is reduced to 1 m^3/d.

6.3 Benchmark Design

The main elements of the BSMs are the simulation models, the simulation protocol, and the performance evaluation criteria. The BSM models are available in different simulation platforms, but the authors ensured that simulations are comparable regardless of the software being used [1]. This section focuses on the design of the BSM1 and BSM2 platforms for MATLAB®. Simulink® models are used to represent the WWTP structure in BSM1 (ASP) and BSM2 (water and sludge lines) platforms.

BSM1 Platform: Three Simulink® models are available:

- *openloop.mdl* for open-loop simulations.
- *benchmarkss.mdl* for closed-loop simulations with ideal measurement signals.
- *benchmark.mdl* for closed-loop simulations with delays and measurement noise.

The BSM1 Simulink® diagram represents the different process units, the stream mixing and splitting points, the controllers, sensors, actuators, and process delays. The models are written in C programming language and incorporated into Simulink® as C MEX S-functions. These C files are predefined for BSMs and should not be modified when BSMs are used for objective comparisons of control strategies. For instance, the C language codes represent the ASM1 model on each reactor block (*asm1*), the 10-layer clarifier model (*settler1dv4*), and the delay block model (*hyddelayv3*) that is introduced to avoid algebraic loops on simulations. The influent is represented using different data files for constant input, dry weather, storm weather, and rain weather conditions (*CONSTINFLUENT, DRYINFLUENT, RAININFLUENT* or *STORMINFLUENT*). The data is in the following order: t, S_I, S_S, X_I, X_S, $X_{B,H}$, $X_{B,A}$, X_P, S_O, S_{NO}, S_{NH}, S_{ND}, X_{ND}, S_{ALK}, Q_{in}. The dynamic influent files are designed for a simulation time of 14 days with a sampling time of 15 min, the first 7 days consisting of dry weather data.

BSM2 Platform: Five Simulink® models are available. The most important are

- *bsm2_ol.mdl* for open-loop simulations and dynamic influent.
- *bsm2_ss.mdl* for simulations with constant influent data.
- *bsm2_cl.mdl* for closed-loop simulations with delays and measurement noise.

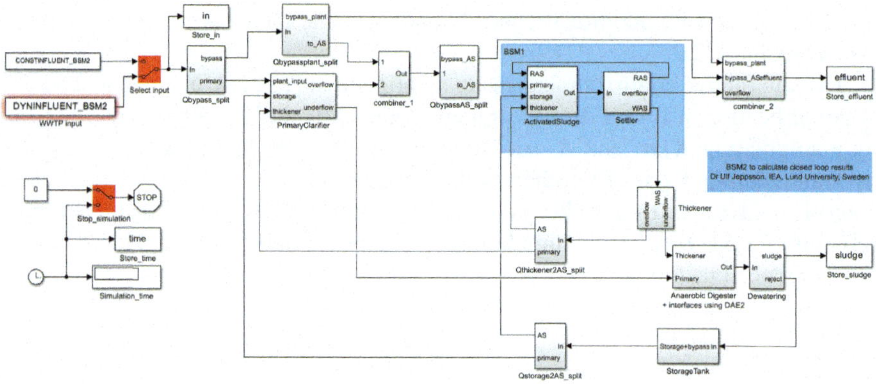

Fig. 6.3 Simulink® model for BSM2 model *bsm2cl.mdl*, including BSM1 highlighted in blue

The two additional models *sensor_actuators.mdl* and *adm1_versions.mdl* are libraries that contain useful models provided by the benchmark developers.

The BSM2 Simulink® model, where ASP is represented by BSM1, is shown in Fig. 6.3. BSM1 reactors receive the input flow from primary settler. The sludge line includes the anaerobic digestion unit represented by ADM1 model; it was necessary to develop an interface (Anaerobic Digester+Interface DAE2 block) to convert variables from ASM1 model (water line) to ADM1 [1, 9]. There are 25 files written in C language for the models of the different units of the plant: ASP reactors, anaerobic digester, stream combiners, delays, stream splitter, primary settler, dewatering, thickener, storage tank, and for solvers for particular models. The data files for dynamic and constant influent (*DYNINFLUENT_BSM2, CONSTINFLUENT_BSM2*) contain the BSM1 influent variables, the influent temperature, and 5 columns for dummy variables (3 soluble and 2 particulate components). The data describe an operation period of 609 days with a sampling time of 15 min. A noise file is provided to approximate real characteristics of sensors and actuators; users can change noise source or choose ideal sensors and actuators [4].

Simulation Protocol: It is necessary to compile the C language files in MATLAB® using the *mexall* and *mexall_bsm2.m* scripts available for BSM1 and BSM2 respectively. In BSM1 platform, the commands *openloopinit* for *openloop.mdl* and *benchmarkinit* for *benchmark.mdl* initiate all variables and parameters and load the data files. In BSM2 platform the initializing command is *init_bsm2*. The ode15s solver can be used for open-loop or constant influent simulations, but ode45 is recommended for closed-loop simulations with dynamic influent. The desired influent file should be selected using the manual switch of Simulink® model (Fig. 6.3). It is recommended to find an initial steady state to start dynamic simulations. In BSM1, it is obtained using *openloop.mdl, benchmarkss.mdl* for simulating 100–150 days, then, *benchmark.mdl* is run for a maximum of 14 days to evaluate ASP closed-loop behavior. In BSM2 platform, 500 days of simulation of *bsm2ss.mdl* are needed to attain steady state. Then, *benchmark.mdl* is used for the long-term closed-loop response (609 days). For dynamic open-loop analysis, *openloop.mdl* (BSM1) and *bsm2_ol.mdl* (BSM2) should be employed choosing the dynamic influent file. The output data (on workspace) represent the dynamic evolution of the variables on each unit of the plant, as well as control variables. The BSMs were originally provided for MATLAB® R2007a/b, but they work in newer versions (actually in MATLAB® R2024a). In recent versions, there might be problems with the consistency of dimensions of the vectors K_La_i and carbon dosage when running performance evaluation codes, and some manual adjustments could be necessary to avoid MATLAB® errors.

Performance Evaluation: The BSMs provide a performance assessment framework divided into two levels. The first level uses standard metrics of control performance [4] as control error, integral of absolute error (IAE), integral of square error (ISE), maximum deviation from setpoint, variance of the controlled variable, and maximum deviation of manipulated variable to asses control system efficacy. The second level measures the effect of the control strategy on plant performance based on indicators of environmental and operational costs. The evaluation period proposed on BSM1 is one week, starting on the 7th day of simulation. The evaluation period on BSM2 is one year, starting on the 245th day of simulation that is July 1st.

BSM1 provides the file *perf_controller* for the first level evaluation of performance. It computes the control metrics in terms of S_O in reactor 5 and S_{NO} in reactor 2, and shows the evolution of the controlled and manipulated variables. BSM2 provides a similar file, *perf_controller_bsm2*, that focuses on DO concentration in reactor 4 and purge flow (Q_w).

For the second level evaluation, indicators of pollutant removal efficacy and metrics of environmental and operational costs are computed using the dynamic output data. The effluent quality indicators and their limits are presented on Table 6.6. The files *perf_plant* and *perf_plant_bsm2* on BSM1 and BSM2 respectively determine the values of Table 6.6 indicators as a function of ASM1 and ADM1 variables, and the number and magnitude of violations of the desired limits and the percentage of time that the plant is in violation [1]. Other indicators of performance are the IQI and EQI that measure the amount of pollutants in the influent and effluent respectively, and OCI index that measures the operating costs including aeration energy, pumping energy, sludge production to be disposed, consumption of external carbon source and mixing energy. In BSM2, the operation costs include the heating energy and the electricity recovered from biogas. The energy content of biogas is estimated as 13.89

Table 6.6 Effluent quality limits

Variable	Value
Total nitrogen, N_{tot}	<18 g N/m^3
Total Chemical Oxygen Demand, COD_t	<100 g COD/m^3
Ammonium, S_{NH}	<4 g N/m^3
Total Suspended Solids, TSS	<30 g SS/m^3
Biological Oxygen Demand, BOD_5	<10 g BOD/m^3

kWh/kg CH_4 and a biogas to electricity efficiency of 43% and a biogas to heat efficiency of 50% is considered in BSM2 performance evaluation [12]. Regarding other properties of the resulting control system such as guaranteed stability and robustness, they are not usually studied in detail for BSMs, because the application of typical analysis methodologies is cumbersome due to the complexity of the processes. In fact, most of the research focuses on the performance indicators described above.

Finally, it is important to mention that there are some phenomena that may cause operational problems of qualitative nature which are not currently considered in BSM1 and its extensions. Then, a risk assessment model for BSM2 is provided in folder *risk_calculation* to evaluate plant reliability. It computes performance criteria related with potential settling problems such as filamentous bulking sludge due to nutrient deficiency, risk of foaming due to low F/M ratio, risk of foaming associated with high readily biodegradable organic matter fraction, and risk of rising sludge.

6.4 Accessing the Benchmark

The presented benchmark is available for all users, including different ASM models. The benchmark was implemented first in general languages FORTRAN, C and MATLAB® and also in commercial WWTP software packages GPS-X, SIMBA, and WEST. In addition, results were validated using other software platforms [1]. Nowadays, MATLAB® is the mainstream benchmark platform for either adding new features or working with well-established versions, though other implementations can be found in other programming languages.

Current version and technical reports are available for download from open access repositories.[1] In particular, it is possible to download BSM1 MATLAB® code with implementations of ASM1, ASM2d, and ASM3 activated sludge models, together with an extension with pH calculations; BSM2 code with some modifications such as the consideration of a reactive settler, or greenhouse gas emissions (GHG); different versions of the anaerobic digestion model ADM1 (with algebraic solutions for H_2 and pH; ionic corrections and others); influent generators for BSM2 and other related tools. Together with MATLAB® code distributions, influent typical data are available for different weather conditions (dry, storm, and rain) for BSM1, and also an extended dynamic influent for BSM2, including temperature seasonal variations.

For this benchmark, no online execution is available, because it is straightforward to download MATLAB® files for Windows, MacOS, or Linux Operating Systems. The only requisite is to mention Associate Professor Dr. Ulf Jeppsson from Lund University (Sweden) as coordinator of the working group in the Acknowledgements section in all research published using this benchmark. The benchmark user is free to modify the models and tools. However, such modifications should be mentioned and documented properly when publishing research results in order to allow fair

[1] Benchmark files and technical reports are available in https://github.com/wwtmodels and https://wwtmodels.pubpub.org/str, respectively.

comparisons [1]. The available documentation and software include the full set of model equations, typical parameter values, influent data, and a procedure to test the code and simulation results. Performance assessment is also standardized and all indexes are described in the technical reports, though new indexes are also used in current research in order to assess new environmental issues.

6.5 Discussion for Future Comparison

In this section, the current work lines concerning the BSMs are presented. It is not possible to describe all the existing proposals to control this benchmark, therefore only the proposals that have a greater connection with control engineering and the most ground-breaking approaches are presented. To mention some of them, there are MPC controllers (linear, nonlinear, economic, or robust), hierarchical and cascade controllers, adaptive optimal controllers, fault-tolerant controllers, together with detailed sensitivity analysis and the inclusion of observers. Each of them has its own strengths depending on the research objectives (performance, costs, optimal operation, fulfillment of effluent legal limits, energy optimization), always showing a trade-off between the various criteria. In Fig. 6.4 a DO control example using the BSM2 default strategy is presented. The aspects that focus the attention of current approaches are the consideration of multiple decision criteria, the integration of subsystems, and the vision of the water treatment cycle from a more holistic perspective, all of them including sustainability concerns.

Benchmarking and Performance Evaluation: The different BSM tools have been crucial for the development of increasingly better control and operation strategies for WWTPs. However, they do not yet include all the features that would be desirable to fully evaluate control strategies. Thus, in [13] a series of extensions of BSM1 and BSM2 platforms are introduced, including new models for the elimination of nitrogen, biological elimination of phosphorus, membrane reactors, the inclusion of greenhouse gas (GHG) emissions, and the addition of new indices to evaluate energy production/consumption. It is worth highlighting the considerations about possible failures in measurements and their quality. The use of the different online measurements available is essential for the implementation of optimal operating strategies for a WWTP. However, it is necessary to state methods to verify the quality of these signals (calibration, drifts, noise). Defective measurements can cause deviations in the effluent, ending in violations of legal limits or increasing operating costs. In this sense, the benchmark platforms can be used to show the validity of strategies and algorithms proposed in the presence of these failures and evaluate their impact on the overall functioning of the plant.

Life Cycle Analysis and Environmental Impact Evaluation: Life Cycle Analysis (LCA) is a methodology used to quantify the impacts associated with all stages of a product, from its creation to its destruction. These environmental impacts are (a) impacts on human health (climate change, ozone depletion, toxicity); (b) impacts on ecosystems (acidification, eutrophication, toxicity), and (c) impacts on resource

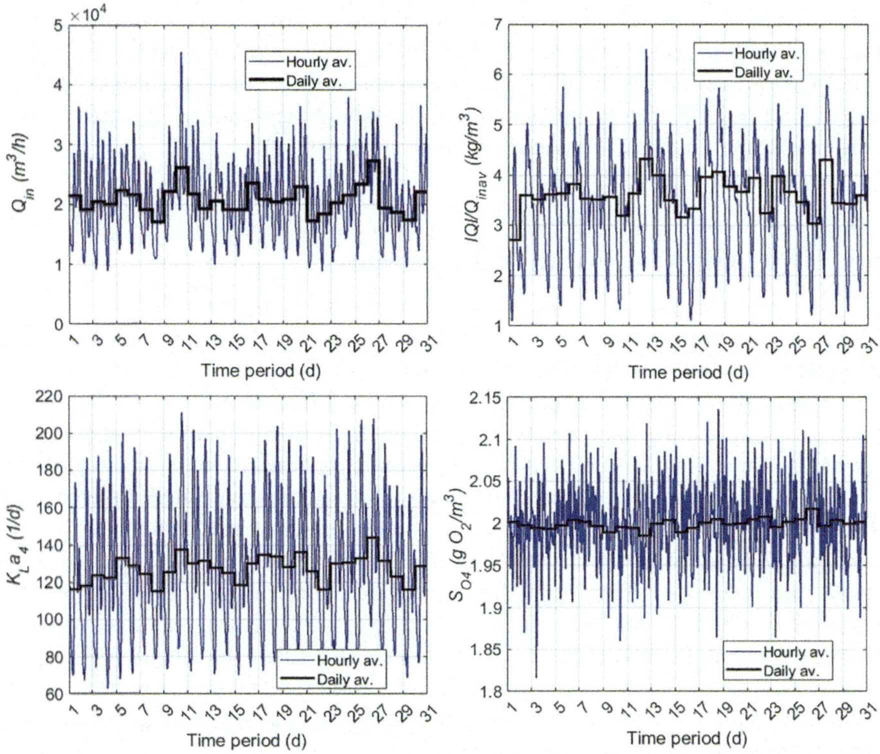

Fig. 6.4 Influent variables and plant response for BSM2 default control strategy

availability (mineral depletion, fuel based on fossil sources). The application of LCA to WWTPs began at the beginning of the 90 s, showing that it is a useful tool to evaluate the environmental impacts of the WWTP operation from a broader perspective. Generally, the LCA methodology serves as a complement within a multi-criteria decision scenario. For example, in [15] LCA is applied to analyze the impacts generated by a set of WWTP operation strategies focused on the BSM1. LCA is proposed as a tool to help in the evaluation of operation strategies, showing how it can complement the quality and cost indices established in the BSM platforms. Together with the introduction of GHG emissions, the application of the LCA shows the interest in incorporating environmental considerations into the evaluation of process operations.

Greenhouse Gas (GHG) Emission Effects: It is important to consider GHG emissions in WWTP operation, since it involves different sources of GHG, including biological treatment, sludge digestion and disposal, and use of energy and chemicals. Among the contributors to GHG, N_2O is produced in the nitrification and denitrification stages of WWTPs, and in the processes of nitrogen and CH_4 elimination in sludge digestion. Moreover, energy used in aeration, pumping, and mixing can be converted into CO_2 equivalents and CH_4 is produced in anaerobic digestion, even if it is exploited for energy recovering. In [14], a model that constitutes the basis for the incorporation of GHGs in benchmarking platforms is proposed. Specifically, it is incorporated into the BSM2, giving rise to the BSM2G benchmark. From the point of view of the BSMs platforms, the purpose is to bring together the different emissions into a new index that can be interpreted as a carbon footprint.

Modeling and Control of the Integrated Urban Water System: The importance of considering the Urban Water System (UWS) from a holistic perspective, including the sewer and receiving river basin, has been widely recognized [13]. The stricter regulations imposed by European directives to ensure a healthy state of the receiving waters have conditioned the UWS management, thus renewing interest in integrated control. Local control focuses only on optimizing performance of the WWTP individual subsystems, but it may not necessarily lead to improve the quality of the receiving waters. In this way, integrated control strategies are necessary to attain global objectives. Moreover, integration is also needed for the transformation of the whole WWTP, where there are energy-consuming and energy-producing elements, in an energy producer instead of consumer. In order to contribute to the development of integrated control methodologies, the availability of benchmarking tools is a decisive factor. In this regard, a first benchmark proposal for UWS has been presented in [13], and a specific benchmark for the sewer is introduced in [16]. The latter is closely related to BSMs because its implementation, variables considered, and performance evaluation are compatible, allowing for their interconnection [17].

Machine Learning and Data-Driven Approaches: Machine learning and data-driven techniques have proven to be a very powerful tool that is revolutionizing industry, specifically the area of intelligent control when using knowledge extracted from data for supervision and control of complex processes, with many successful practical applications. Therefore, their application to WWTP and the corresponding benchmark BSM1 has been a very active research field in the last years because plant sensors generate a large amount of data, and those large datasets are likely to be used for monitoring, performance analysis, prediction, or data-driven control [18].

Acknowledgements To IWA Task Group, IEA Department, Lund University, Sweden (U.Jeppsson, C. Rosen), for the development and free distribution of benchmarks BSMs. To Spanish Government: PID2019-105434RB-C31, PID2019-105434RB-C33, TED2021-129201B-I00 and MICINNU/AEI/10.13039/501100011033 projects.

References

1. Gernaey K, Jeppsson U, Vanrolleghem P, Copp, J (2014) Benchmarking of control strategies for wastewater treatment plants. Scientific and technical report No. 23. IWA publishing
2. Santín I, Pedret C, Vilanova R (2016) Control and decision strategies in wastewater treatment plants for operation improvement. In: Intelligent systems, control and automation: science and engineering. Springer
3. Pons MN, Alex J, Benedetti L, Copp JB, Gernaey K, Jeppsson U, Vanrolleghem P (2021) Benchmark simulation model no. 1 (BSM1). Wastewater modelling. https://wwtmodels.pubpub.org/pub/web0fxkd

4. Alex J, Benedetti L, Copp J, Gernaey K, Jeppsson U, Nopens I, Vanrolleghem P (2021) Benchmark simulation model no. 2 (BSM2). Wastewater modelling. https://wwtmodels.pubpub.org/pub/ytwymnp3
5. Vrecko D, Gernaey K, Rosen C, Jeppsson U (2006) Benchmark simulation model No 2 in Matlab-Simulink: towards plant-wide WWTP control strategy evaluation. Water Sci Technol 84:65–72
6. Henze M, Gujer W, Mino T, Van Loosedrecht M (2006) Activated sludge models ASM1, ASM2, ASM2d and ASM3. IWA Publishing
7. Flores-Alsina X, Gernaey K, Jeppsson U (2012) Benchmarking biological nutrient removal in wastewater treatment plants: influence of mathematical model assumptions. Water Sci Technol 65(8):1496–1505
8. Batstone D, Keller J, Angelidaki I, Kalyuzhnyi S, Pavlostathis S, Rozzi A, Vavilin V (2002) The IWA anaerobic digestion model no 1 (ADM1). Water Sci Technol 45(10)
9. Nopens I, Batstone D, Copp J, Jeppsson U, Volcke E, Alex J, Vanrolleghem P (2009) An ASM/ADM model interface for dynamic plant-wide simulation. Water Res 43(7):1913–1923
10. Jeppsson U, Pons M, Nopens I, Alex J, Copp JB, Gernaey K, Rosen C, Steyer J, Vanrolleghem P (2007) Benchmark simulation model No 2: general protocol and exploratory case studies. Water Sci Technol 56:67–78
11. Gernaey K, Rosen C, Benedetti L, Jeppsson U (2009) BSM2: a model for influent data generation. https://wwtmodels.pubpub.org/pub/web0fxkd
12. Nopens I, Benedetti L, Jeppsson U, Pons M, Alex J, Copp JB, Gernaey K, Rosen C, Steyer J, Vanrolleghem P (2010) Benchmark simulation model No 2: finalisation of plant layout and default control strategy. Water Sci Technol 62:1967–1974
13. Jeppsson U (2017) The benchmark simulation modelling platform—areas of recent development and extension. In: Lecture notes in civil engineering, pp 81–91
14. Flores-Alsina X, Corominas L, Snip L, Vanrolleghem P (2011) Including greenhouse gas emissions during benchmarking of wastewater treatment plant control strategies. Water Res 45(16):4700–4710
15. Meneses M, Concepción H, Vrecko D, Vilanova R (2015) Life cycle assessment as an environmental evaluation tool for control strategies in wastewater treatment plants. J Clean Prod 107:653–661
16. Saagi R, Flores-Alsina X, Fu G, Butler D, Gernaey KV, Jeppsson U (2016) Catchment & sewer network simulation model to benchmark control strategies within urban wastewater systems. Environ Model Softw 78:16–30
17. Vasiliev I, Luca L, Barbu M, Vilanova R, Caraman S (2024) Reducing the environmental impact of sewer network overflows using model predictive control strategy. Water Resour Res 60:e2023WR035448
18. Pisa I, Morell A, Vicario JL, Vilanova R (2023) Transfer learning in wastewater treatment plants control: measuring the transfer suitability. J Process Control 124:36–53

Chapter 7
On a Benchmark Problem for Automated Insulin Delivery in Type 1 Diabetes

Vihangkumar V. Naik, Clara Escorihuela-Altaba, and Jose Garcia-Tirado

7.1 Introduction

7.1.1 Background

Type 1 Diabetes (T1D) is an autoimmune condition of difficult management resulting from a progressive loss of insulin-producing beta cell mass with an increased risk of acute and long-term macro- and microvascular complications. Epidemiological data indicate that people with T1D have lifespans $\sim$11–13 years shorter and experience cardiovascular disease events on average more than a decade earlier than the general population. Exogenous insulin replacement is the only known effective therapy to treat this condition.

1993 marked the landmark Diabetes Control and Complications Trial (DCCT) where the concept of Intensive Insulin Therapy (IIT) was introduced [7]. IIT consisted of a combination of basal and prandial insulin injections throughout the day combined with multiple self-monitoring blood glucose tests via a glucose meter. IIT effectively reduced average glycemia and microvascular complications. However, IIT is often associated with increased time in hypoglycemia (low glucose levels, i.e., $<$70 mg/dL or 3.9 mmol/L), which can be associated with serious complications and even death if untreated. Conversely, prevailing exposure to hyperglycemia (high glucose levels, i.e., $>$180 mg/dL or 10 mmol/L) has serious short- and long-term implications on

V. V. Naik · C. Escorihuela-Altaba · J. Garcia-Tirado (✉)
University Clinic for Diabetes, Endocrinology, Nutritional Medicine, and Metabolism, Inselspital-University Hospital Bern, University of Bern, Bern, Switzerland
e-mail: jose.garcia@unibe.ch

V. V. Naik
e-mail: vihangkumar.naik@unibe.ch

C. Escorihuela-Altaba
e-mail: clara.escorihuelaaltaba@students.unibe.ch

J. M. Maestre and C. Ocampo-Martinez (eds.), *Control Systems Benchmarks*, Advances in Industrial Control, https://doi.org/10.1007/978-3-031-76312-0_7

health and life expectancy. Hence, people with T1D face a non-ending challenge of controlling their blood glucose levels within a safe range (70–180 mg/dL) while minimizing exposure to hypo- and hyperglycemia.

T1D is a condition well known to impose a high economic, physical, social, and mental impact on individuals and their caregivers [9]. In addition, healthcare systems all around the world face an ever-increasing economic burden, projected to be near US$ 2.2 trillion by 2030 [3].

According to recent figures [9], the number of individuals living with T1D is estimated at around 8.4 million globally, from which 1.5 million are < 20 years. An increase of 0.5 million new cases was diagnosed and $\sim$175,000 individuals with T1D died in 2022, from which 35,000 were due to lack of diagnosis. Projected numbers for 2040 indicate a substantial increase in the prevalence of T1D, with estimates ranging from 13.5 to 17.4 million cases, representing an increase of 60% to 107% compared to 2021. Notably, this increase is expected to be prominent in low-income and lower-middle-income countries, presenting the need for preventive measures or, after the onset, affordable solutions for disease management.

7.1.2 Current Therapeutic Alternatives in T1D

Existing treatments for T1D aim at maximizing the time spent in the normoglycemic range (percent time spent between 70–180 mg/dL (TIR)) while minimizing the percent time spent below 70 mg/dL (TBR) and above 180 mg/dL (TAR) through an orchestrated combination of diet, glucose self-monitoring, and exogenous insulin administration. The administration of external insulin is broadly categorized into Multiple Daily Injections (MDI) and Continuous Subcutaneous Insulin Injections (CSII). The latter can be configured with open- and closed-loop strategies (a.k.a. the Artificial Pancreas).

In the MDI approach, two types of insulin are administered at least once a day: basal (long-lasting) and prandial/bolus (fast-acting) insulin via disposable pens equipped with pre-filled or refillable insulin cartridges. Basal insulin covers the insulin needs between meals and overnight, whereas prandial/bolus insulin aims to counteract the postprandial excursion after carbohydrate ingestion. Long and fast-acting insulin doses are calculated based on an individualized set of parameters determined by the treating physician, including basal rate (BR), which determines the daily dose of long-acting insulin, carbohydrate-to-insulin ratio (CR), a parameter reflecting the insulin dose necessary to cover a specific quantity of carbohydrate intake, and the Sensitivity Factor (SF) that indicates the drop in blood glucose caused by one unit of insulin.

In the open-loop CSII, the insulin pump is programmed with a predefined basal rate profile (BR) over the day. To manage the postprandial glycemic elevation, the user manually programs the prandial injection as a function of the CR and SF. On the other hand, the closed-loop control (CLC) system attempts to mimic the normal pancreas function. As shown in Fig. 7.1, CLC consists of a Continuous Glucose

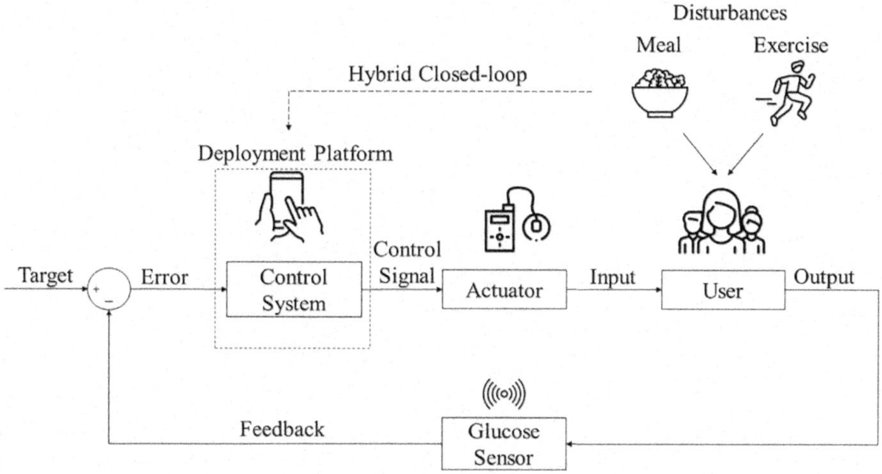

Fig. 7.1 Basic blocks of an CLC system

Monitoring (CGM) (sensor), an insulin pump (actuator), and a control algorithm. Based on the CGM measurements, the control system modulates the BR in real time to keep the blood glucose (BG) in the target range.

In terms of automation, there are two main CLC modalities: (i) hybrid CLC (hCLC/hAID) and (ii) fully automated CLC (fCLC/fAID). (hCLC)s automate the pump's BR but (1) do not automate the prandial boluses and (2) are not well suited to avoid hypoglycemia during and after physical activity. In this sense, the users must interact with the app/pump and provide explicit information to counteract these two main glycemic disturbances. Regarding meals, the user must manually enter information about the carbohydrate content to obtain a bolus recommendation that needs to be approved before the injection. In the case of physical exercise, the user must program the physical activity to allow for proper insulin reduction/suspension to protect against hypoglycemia. This approach may include a second type of boluses (correction boluses), administered to the user when the basal profile is insufficient to drive the CGM back to the target range. Unlike hCLC, fCLC aims at removing the user from the loop, automating the prandial bolus and insulin regime in response to physical activity.

Figure 7.2 shows a simulation example of each treatment procedure given a CGM profile for an illustration purpose. Here, the MDI approach administers a single basal dose per day. In open-loop CSII, the physician specifies a unit/hour insulin ratio to determine the BR. In Automated Insulin Delivery (AID), the control system automates the BR based on CGM readings. The latter approach also includes a correction bolus if the basal dose is insufficient to maintain the CGM in the target range. All treatments must be announced regarding meal and exercise events to calculate the prandial bolus (based on the CR and SF) and the insulin reduction/suspension due to exercise.

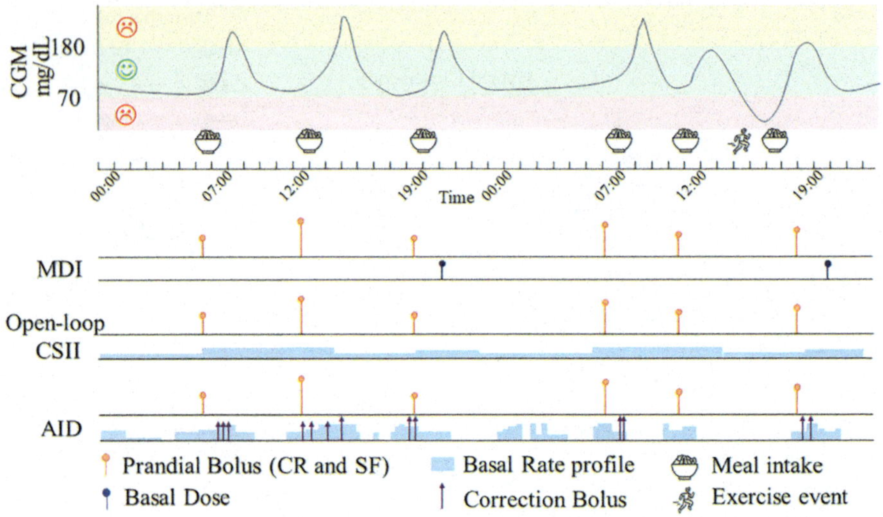

Fig. 7.2 An illustration of the therapeutic alternatives in T1D

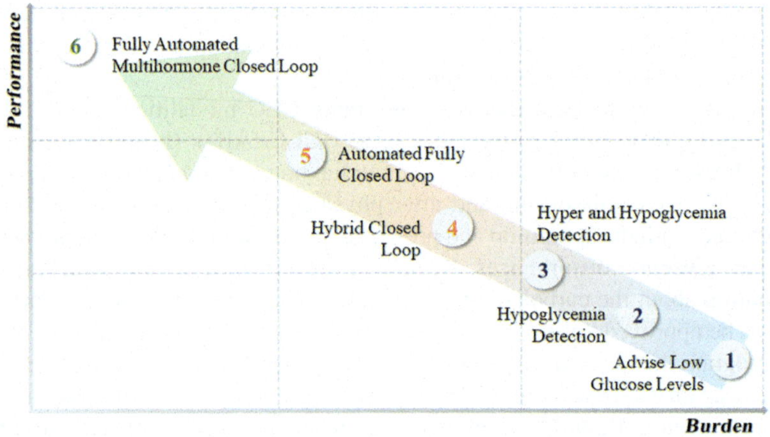

Fig. 7.3 AP levels of subject burden and system performance in terms of glucose control

In 2006, the Juvenile Diabetes Research Foundation International (JDRF) launched the Artificial Pancreas (AP) Project to accelerate the development of commercially viable AP systems and in 2009, the Project was envisioned as a 6-step development pathway [13]. The AP levels for subject burden and system performance are depicted in Fig. 7.3. Starting from an insulin pump designed to minimize the risk of hypoglycemia, the development progresses toward a fully automated CLC system. This advanced system is meant to autonomously adjust insulin doses, including BR, prandial (meal) and correction boluses (for high blood sugar), in response to (or prediction of) meals and/or physical activity. This aims at decreasing the disease-related burden while targeting optimized glucose management.

Nowadays, hCLC is the most advanced treatment for T1D, providing better glucose control and less diabetes distress compared to other alternatives. According to Nwokolo and Hovorka [11], research on exploring alternative inputs and algorithms to detect and predict the occurrence of meals and physical activity and designing dual hormone closed-loop approaches (third-generation) is ongoing. However, fCLC systems have not yet reached a commercial stage. Section 7.1.4 aims to highlight the challenges and draw the path toward designing and evaluating a second generation of fCLC systems.

7.1.3 Existing CLC Algorithms

In terms of used control system algorithms and methodologies, model predictive control (MPC) dominates the landscape due to its flexibility and capacity to handle constraints directly. However, proportional–integral–derivative (PID) and fuzzy logic approaches have also been implemented both in research and commercial devices. In September 2016, the Food and Drug Administration (FDA) approved the first hCLC system (the Medtronic 670G, now replaced by the 780G model), which is capable of automatically adjusting BR but does not automate prandial insulin boluses. To date, four AID systems are cleared/approved for use in the US together with the Medtronic 780G: t:slim X2 with Control-IQ (Tandem), Omnipod 5 (Insulet), and iLet (Beta Bionics). Tidepool Loop was recently approved in the US as an interoperable Automated Glycemic Controller (iAGC). Two more systems are available in Europe: Ypsomed's CamAPS FX and Diabeloop's DBLG. None of them can automate prandial boluses nor predict physical activity.

7.1.4 Control Challenges in AID

As presented below, the control system design for AID systems poses many practical challenges related to sensor and actuator delays, uncertain mathematical models (large inter-patient and intra-patient variability), unidirectional control, and unmeasured disturbances. Unidirectional control action arises in this problem since insulin decreases glucose levels, not vice versa; the pump can be shut when glucose levels are low, but the user might need a hypoglycemic treatment (fast-acting carbohydrates) to recover.

Limitations of Current Continuous Glucose Monitors: A CGM device is typically composed of a sensor (placed under the skin), a (Bluetooth) transmitter, and a receiver. Due to the sensor location, glucose levels are measured in the Interstitial Fluid (ISF) and not directly in the blood. As a result, rapid changes in glucose levels are observed in the ISF only after the diffusion of glucose molecules through the capillary vessels, which might typically take ~10–15 min on average. Recent devices addressed the ISF-to-plasma delay with estimators embedded in the post-processing of the raw measurements, which can compensate for delays in the near steady state. However, compensation for rapid glucose changes (transient state) from meals and physical activity is still an open problem.

Current Insulin Preparations: The ability to respond to a rapid increase in glucose in the early postprandial state is often limited by available rapid-acting insulin analogs due to their inherent absorption delay. From a control systems standpoint, shortening such an absorption delay would be a natural and straightforward way to improve the performance of AID systems. Ultra-rapid insulin formulations are a new class of prandial insulin analogs showing earlier insulin exposure (peak action) than standard insulin preparations. Although only limited data has been published, the early indication is that unmodified control strategies do not lead to satisfactory glycemic control. Faster formulations and advances in control designs are likely necessary to enable full prandial automation.

Challenging Postprandial Glucose Control: Blood glucose concentration in the postprandial state (after meal ingestion) reflects a net balance between the amount of glucose entering (Rate of Appearance: R_a) and leaving (Rate of Disappearance: R_d) the circulation. R_a largely depends on the rate of appearance of meal carbohydrate (MR_a) and the rate of Endogenous Glucose Production (EGP) while R_d reflects the rate of whole-body glucose uptake (largely dependent on insulin action). The above is due to the limitation of current pharmacokinetics and pharmacodynamics (PK/PD) of modern insulin analogs, which still fail to provide an exact match between (R_a) and (R_d) to and from the circulation, respectively, leading to significant postprandial excursions.

Timely Recognition of the Postprandial State: Meal detection has attracted the attention of the diabetes technology community for a long time, aiming at its integration into AID systems. Methods based on CGM have led to detection times of ~30–40 min when the hyperglycemic excursion is challenging to avoid. Alternative methods other than those using glucose measurements or insulin–glucose models have been developed using chewing movement, sound, images, and accelerometry; most of these techniques report detection rates around 90% but with much shorter detection lags than glucose-based methods. None of these faster methods have yet been integrated into diabetes care. An important caveat of all these methods is that none can accurately estimate the amplitude of a meal, at least not before significant time (hours) elapses, and even then, with limited accuracy. Therefore, these methods cannot alone replace user input in a meal bolus strategy, even in hCLC designs.

Main Glycemic Disturbances: Meals and physical activity constitute the two primary glycemic disturbances in T1D. In hCLC, the correct estimation of the carbohydrate content is essential because it determines the size of the prandial bolus to be administered. Since the glucose appearance in the bloodstream after a meal is faster than the rate of insulin absorption in the subcutaneous route, injection timing is also paramount (recommended 10–15 min before the first bite). On the other hand, physical activity may induce a glucose-lowering or glucose-increasing effect, depending on the duration and intensity. Light-to-moderate-intensity aerobic exercise often causes blood glucose concentrations to drop rapidly, while resistance exercise may increase glucose levels. While High-intensity interval training (HIIT) can balance blood sugar levels, adherence to this activity type is difficult to achieve [21]. This heterogeneity in glucose response to different types of physical activities makes glucose control especially challenging.

7.2 Problem Description

This section briefly describes and compares the physiology of glucose homeostasis in individuals with and without diabetes and presents the current mathematical models for simulation. Furthermore, this section also outlines the challenges encountered in modeling such a complex process. The models presented have been used for various applications, including but not limited to in silico testing for novel AID algorithms, learning and understanding of diabetes dynamics, and prediction assessment.

7.2.1 Glucose Homeostasis: Diabetes Versus Healthy Body

To address the challenges outlined in the former section, it is essential to account for the underlying physiological knowledge of insulin and glucose metabolism as the foundation of mathematical models for simulation and control system design.

Glucose is the primary source of energy in the human body. Maintaining optimal blood glucose levels is essential for ensuring a continuous energy supply. In healthy individuals, this is achieved by a process known as glucose homeostasis; see the graphical representation in Fig. 7.4. The stomach and intestines (gastrointestinal tract) digest the ingested carbohydrates into glucose molecules to be absorbed into the bloodstream. The incretin effect signals the pancreas via the glucagon-like peptide-1 (GLP-1) and Glucose-dependent Insulinotropic Polypeptide (GIP) to produce insulin from the first bite via a well-orchestrated predictive action. Glucose is utilized via direct insulin action, while the remanent is stored as glycogen in the liver. When glucose levels drop, the pancreas releases glucagon, a hormone signaling the liver to break down glycogen into glucose via glycogenolysis.

In individuals with T1D the pancreas cannot produce insulin, requiring exogenous insulin administration via MDI or CSII to ensure the insulin-dependent distribution of glucose to the tissues. Insulin is injected in the subcutaneous space, leading to a delay in the absorption and distribution from the ISF into the bloodstream. Like healthy individuals, glucose is filtered and reabsorbed by the glomeruli in the kidneys. However, kidneys have a maximum reabsorption capacity; once this capacity is reached due to excessive glucose, glucose is eliminated through the urine via glucosuria (renal excretion). Although endogenous glucagon is still produced by the

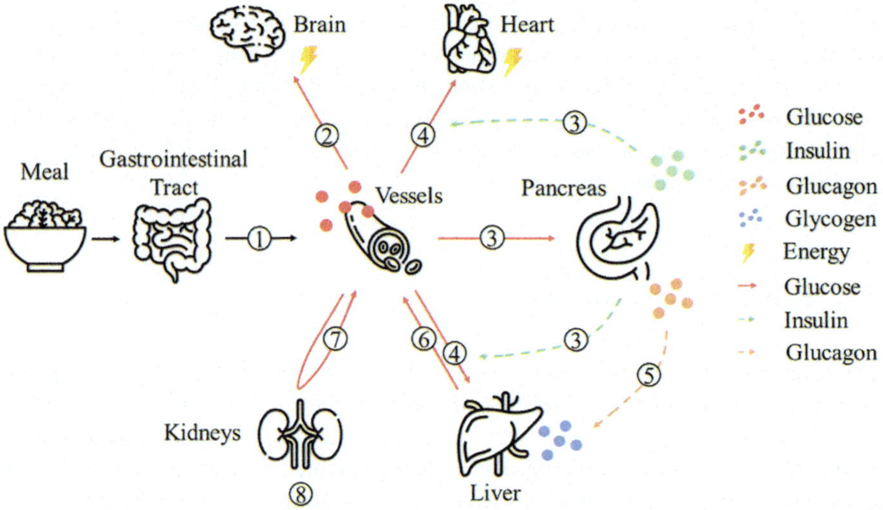

Fig. 7.4 Glucose homeostasis process on a healthy subject: (1) Gastrointestinal organs process carbohydrates and facilitate glucose absorption into the bloodstream. (2) The brain continually consumes glucose independently of insulin. (3) Elevated blood glucose levels trigger an increase in insulin concentration. (4) Glucose distribution to other tissues is insulin-dependent, with muscles utilizing glucose for energy and the liver storing it as glycogen. (5) Pancreatic secretion of glucagon is stimulated when blood glucose levels are low. (6) Glucagon converts glycogen into glucose. (7) Under physiological conditions, glucose is reabsorbed into the bloodstream via kidney function

alpha cells, its role in glucose homeostasis is deflected, therefore losing the ability to recover from hypoglycemia.

7.2.2 Phenomenological/Physiological Models in T1D

Phenomenological/physiological models aim to abstract the available physiological knowledge in T1D using mathematical models of ordinary differential equations. These models are designed to approximate glucose and insulin dynamics in T1D through one or more compartmental submodels. Regarding complexity, two main types of models arise in diabetes technology: minimal and maximal [5]. Minimal models describe key elements of glucose homeostasis and are generally considered for control algorithm synthesis; maximal models are more detailed and include much more information and detail compared to minimal models. These models are typically used in simulation to reproduce realistic scenarios to be tested in clinical trials for a virtual population of T1D individuals. In terms of control systems engineering, the typical inputs are exogenous insulin (rate, controlled variable), meals (rate, disturbance), and physical activity (disturbance); the output is glucose concentration via CGM. In some applications, activity trackers provide additional information to predict changes due to physical activity (heart rate, steps, etc.).

The minimal model was introduced by Bergman in 1979 in an intravenous glucose tolerance test (IVGTT) experiment [2]. This model comprises only two compartments to estimate changes in the glucose and insulin concentrations in plasma. The Sorensen and the UVA/Padova simulator are situated at the other extreme of the spectrum and are therefore considered maximal models. The UVA/Padova simulator was developed by the University of Padova and the University of Virginia's Center for Diabetes Technology in 2008. The U.S. Food and Drug Administration accepted this simulator as a substitute for animal pre-clinical trials, accelerating research by enabling the substitution of expensive pre-clinical trials with thorough and sound simulations of the targeted scenarios. The simulator underwent subsequent modifications in 2014 and 2018, integrating four submodels representing glucose dynamics, insulin administration and distribution, carbohydrate absorption, and glucagon kinetics. It also models phenomena observed in T1D individuals, such as the dawn phenomena and the time-varying insulin sensitivity as a function of the circadian rhythm.

The Sorensen model is a result of a Ph.D. thesis presented in [22], subsequently revised in [17]. This model represents the glucose dynamics in healthy and T1D subjects. The Hovorka model [10], developed at the University of Cambridge, comprises eight differential equations and integrates three subsystems: glucose kinetics, the insulin subsystem, and insulin action. Furthermore, the Hovorka model was combined with a two-compartment meal model describing the carbohydrate absorption and conversion into plasma glucose in [25].

The Medtronic's "Medtronic Virtual Patient (MVP)" [12] features five equations and eight time-invariant identifiable parameters [12], comprising two distinct submodels: one for insulin absorption, and another for glucose absorption after meal ingestion. The subcutaneous oral glucose minimal model (SOGMM) [18] integrates the glucose–insulin core model from Bergman's minimal model with subcutaneous insulin administration and oral carbohydrate absorption. Moreover, there are numerous variations of these models documented in the literature aiming to find a trade-off between model complexity and prediction accuracy.

7.2.3 Challenges of Physiological Modeling

This section presents the common limitations inherent to mathematical models based on differential equations.

Some known scenarios, such as exercise, stress, and illness, primarily affect glucose dynamics. However, their accurate representation using appropriate mathematical equations is challenging. Several attempts to address these challenges have been presented in the literature [5, 23], but resulting in simplifications that do not fully capture the dynamical variations of the underlying physiology.

The challenge to accurately estimating carbohydrate intake input directly affects the glucose submodel for predicting the CGM. This makes it important not only for control but also for modeling perspective to have an accurate computation of CR. Additionally, establishing initial conditions poses difficulties; the more straightforward approach is calculating the steady-state conditions. However, this does not account for potential dependencies on prior glucose dynamics, making the initial conditions equal to any prior scenario.

Another significant challenge is assessing the observability and identifiability of the models to assess the ability to infer the states and parameters from measured outputs. The model parameters must be estimated from empirical data. However, structural and practical identifiability issues may hamper this task. Sometimes, the model structure does not yield a unique parameter solution. Conversely, even when a unique parameter solution emerges from the model structure, a wide range of parameter values may produce similar predictions given the uncertainty from the measured variables (noise). In addition, it is of utmost importance to quantify the inter- and intra-patient variability of the model parameters.

Identifying control-oriented models in diabetes technology has followed a relatively informal approach, with a primary focus on insulin sensitivity. However, this approach does not consider the consequent effect on other parameters and measurements of the system.

7.2.4 Alternative Modeling Approaches

In contrast to physiological models, data-driven models can be useful in capturing complex relationships between inputs and outputs without relying on explicit knowledge of the underlying physical processes. The availability of CGM systems, the increase in medical history collection, and the popularization of wearable devices have generated a large amount of medical data in the field of diabetes. This has significantly driven the development of new algorithms based on artificial intelligence, machine learning, and time series analysis, including ensemble, deep learning, and autoregressive models [23].

In this context, the development of data-driven models is of particular interest, as they could be used to continue or complement the models discussed above. Glucose prediction primarily uses signals recorded before the prediction time, such as CGM, insulin on board, carbohydrate intake, exercise tracking, and emotional cues, to provide glucose predictions up to 60 min in the future. Unlike physiological models, the prediction is based on historical data, capturing the subject's past relationships between CGM and insulin.

Pattern recognition techniques are needed to identify recurring episodes within time series data and to make associations with past and current events. Several attempts have been made to cluster daily profiles. However, clustering of CGM and insulin patterns is still an open problem. This, combined with the integration of other biomedical signals recorded by wearables, has the potential to predict uncertain situations such as food intake, physical activity, illness, and stress in advance. Integrating this information into the model and/or the control algorithm can improve predictions and prevent dangerous situations by adding constraints and adapting actions based on previous information.

More recently, personalization has also benefited from AI. This is mainly achieved by designing neural networks that can (1) fine-tune parameters, (2) mimic the dynamic systems of physiological models (since an Artificial Neural Network (ANN) is easier to personalize to individual data), and (3) integrate phenomenological reasoning into their predictions. These approaches aim to combine phenomenological and data-driven models to take advantage of each technique. A review of data-driven models for diabetes management is detailed in [26].

Although data-driven models for diabetes management are becoming increasingly important [26], it is worth emphasizing that they present their own challenges. When dealing with healthcare data, generalization can be difficult, as each individual may respond uniquely to the same stimulus. They may struggle when dealing with poorly represented datasets, potentially introducing bias when applied to different sample populations. In addition, their use in healthcare raises regulatory concerns, requiring strict adherence to regulatory guidelines, highlighting the need for responsible and ethical use of AI models in health care.

7.3 Benchmark Design

This section describes the tools available to evaluate T1D-related control systems using in silico virtual testing on patients. It also describes the medically based metrics and test scenarios for evaluating the efficacy of the devised control algorithm.

7.3.1 Current Available Simulators

Various research groups have introduced simulators aimed at facilitating the experimentation of the developed control systems (Table 7.1). These simulators predominantly consist of two components: one for simulating virtual patient data, which corresponds to the plant of the control system, and another intended for the integration of the control algorithm. Their combination enables the analysis of prospective glucose levels in response to the control algorithm actions.

As pointed out, the CGM from the virtual patient model is considered a measured output, and the insulin dosages to be administered constitute the control input. This input, in conjunction with the other required inputs from the specific model residing in the plant, affords the simulation model the capacity to forecast the forthcoming trend in the CGM profile of the virtual patient. The resulting CGM profile will subsequently undergo evaluation through the metrics described in Sect. 7.3.3, affording a quantitative assessment of the control system performance.

Table 7.1 T1D Simulator platforms

	Open source	License	Code	Glucodyn. model	Number of virtual patients
T1DMS (S2017)	No	—	MATLAB®	UVA/ Padova 2017	100 adults, 100 adolescents, 100 children
T1DMS (S2008c)	No	Commercial	MATLAB®	UVA/ Padova 2008	10 adults, 10 adolescents, 10 children
Simglucose [27]	Yes	MIT	Python	UVA/ Padova 2008	10 adults, 10 adolescents, 10 children
OHSU [11]	Yes	Free for academic use	MATLAB®	Hovorka	99 sampled from Hovorka parameter distribution
McGill [16]	Yes	MIT	MATLAB®	Bergman	2 adults, original and adapted parameter of Bergam model
				Hovorka	Random adults from Hovorka parameters distribution
GlucoSym [19]	Yes	MIT	Javascript, Pug, CSS. Requires NODE.JS	Medtronic	10 adults
APS TestBed [1]	Yes	GNU	Python, HTML, CSS, JavaScript, Rubz, Mako	UVA/Padova (Simglucose)	10 adults, 10 adolescents, 10 children
				Medtronic (Glucosym)	10 adults
LT1 [14]	Yes	MIT	Vue, Javascript, html	UVA/ Padova	1 adult
				Hovorka	1 adult
CGMsim [4]	Yes	MIT	Javascript, Shell	Combination of previously mentioned models	—

7.3.2 Accessing Different Benchmarks

Table 7.1 presents the specifications of a compilation of eight T1D simulators, which can facilitate evaluating the designed control system. The specifications include important details such as the license terms of the platform, the programming language used in its implementation, the corresponding GitHub repository if the code is available, the model embedded in the plant to simulate the virtual patient data, and the number of virtual patients available for the simulation. Each virtual patient has all the personally calibrated parameters required for the model embedded in the plant by default. Detailed instructions on executing the code and utilizing the benchmark are available within the respective GitHub repositories.

UVA/Padova Proprietary T1D Simulator (S2017): This proprietary simulator is of exclusive use to the original partner institutions, although the model equations are publicly disclosed [24]. This simulator accounts for two main subsystems describing glucose and insulin dynamics. In addition, routes for glucose appearance and disappearance through meals (gastrointestinal tract), EGP by the liver, renal elimination, and insulin-dependent and -independent glucose utilization are added to mimic the pathophysiology of T1D. Insulin appearance is modeled by three different routes of administration: inhaled, intra-dermal, and subcutaneous. This version allows for a thorough testing of insulin-related treatments in 100 virtual subjects in each subpopulation: children, adolescents, and adults.

UVA/Padova Commercial T1D Simulator (S2008c): Commercialized by the Epsilon Group, this simulator gathers the main features of the seminal work proposed by Dalla man et al. [6]. This version allows to experiment with ten virtual subjects in each subpopulation: children, adolescents, and adults.

Simglucose is an open-source simulator based on an adaptation of the UVA/Padova S2008c model implemented in Python, developed by Xie [27] for research purposes only. It includes models of 30 virtual patients comprising groups and average age of 10 children (9 years), 10 adolescents (16 years), and 10 adults (52 years), allowing tests using single or multiple patients. *Simglucose* is designed for reinforcement learning with a configurable reward function and provides a simple interface to test other control system algorithms.

The *OHSU* platform, developed by the Artificial Intelligence for Medical Systems (AIMS) lab at the Oregon Health Science University, provides virtual patients with the possibility of using both single- and dual-hormone systems (insulin only or insulin and glucagon) integrated with an additional aerobic exercise model [11]. The single hormone virtual patient is based on the Hovorka model, and the dual hormone virtual patient model additionally incorporates two compartments for glucagon kinetics and glucagon dynamics model, respectively. The population of 99 virtual patients with an average weight of 76 kg and age of 35 years is generated with varying degrees of insulin and glucagon sensitivity sampled from a normal distribution [20].

The *McGill* simulation environment for AP in T1D was designed by the McGill Artificial Pancreas Lab team [8, 16]. It incorporates the Bergman and Hovorka virtual patients with configurable exercise plans. The available controller includes MDI therapy, MPC controller, PID controller, and an open-loop pump therapy. This simulator allows single and multi-patient simulations.

GlucoSym Simulator: GlucoSym was developed by Perceptus.org, a non-profit organization dedicated to developing open-source tools for the diabetes community. *GlucoSym* is specifically designed to test and develop AID systems by using predictions from the Medtronic virtual patient to replicate some of the glucose dynamics observed in different clinical studies [19]. This simulator allows the use of up to 10 different T1D patient models, identified using data collected during a clinical trial with adults (avg. 42 yo; avg. diabetes duration 18 y) [12]. *GlucoSym* allows testing different proposed control systems either individually or in the provided population.

APS TestBed, developed by the University of Virginia's Dependable Systems and Analytics Group (UVA-DSA), consists of a platform that incorporates *UVA/Padova* and *Simglucose*, and therefore the characteristics of each [1, 28]. *APS TestBed* has an adverse event simulation module to test the AID against common events such as hypo- and hyperglycemic events and malfunctions of devices such as CGM sensors, insulin pumps, or controllers. Such faults are applied by employing disturbances into the input–output model of the controller. It also provides an interface to load and test external control algorithms.

The *LoopInsightT1 (LT1)* is a web-based open-source simulator resulting from a collaboration between the University of Applied Sciences in Ulm (Germany) and the University of Lübeck (Germany) [14]. Its goal is to advance diabetes education and research by mimicking glucose trends using the UVA/Padova and Hovorka models. In terms of AID algorithms, the options implemented in simulators are CSII (Basal-Bolus), Proportional–Integral–Derivative (PID) control, and OpenAPS developed by the Do-It-Yourself (DIY) community.

Finally, *CGMsim*, a project led by Lorenzo Sandini, is an interactive real-time T1D simulator for teaching purposes, constructed using a combination of glucodynamical models data [4]. It can simulate insulin infusion using MDI, CSII, and DIY AP systems such as AndroidAPS and OpenAPS.

7.3.3 Performance Metrics and Test Scenarios

Currently, CGM systems provide a plasma glucose concentration estimate every 5 min, or 288 data points daily, allowing for comprehensive day-by-day actionable information on dynamic blood glucose variation. As presented by Maahs et al., a CGM trace provides more meaningful information to the patients, caregivers, and healthcare providers, showing highs, lows, trends, and the effect of different behaviors when compared with Self-Monitoring Blood Glucose (SMBG) [15]. Bearing this in mind, a group of experts proposed a set of CGM-based metrics to assess the performance of AID systems, including but not limited to the percentages of time in the target range (70–180 mg/dL, TIR), below range ($<$ 70 mg/dL, TBR), and above range ($>$ 180 mg/dL, TAR), among others, to enable personalized treatment plans for

Table 7.2 Basic KPI recommendations for AID Control applications[a]

Sr. No.	Measure
1	Mean CGM glucose
2	% CGM time <50 mg/dL (<2.8 mmol/L)
3	% CGM time <60 mg/dL (<3.3 mmol/L)
4	% CGM time <70 mg/dL (<3.9 mmol/L)
5	% CGM time 70–140 mg/dL (3.9–7.8 mmol/L)
6	% CGM time 70–180 mg/dL (3.9–10.0 mmol/L)
7	% CGM time >180 mg/dL (>10.0 mmol/L)
8	% CGM time >250 mg/dL (>13.9 mmol/L)
9	% CGM time >300 mg/dL (>16.7 mmol/L)
10	CGM Standard Deviation
11	CGM Coefficient of Variation

[a] these are a part of recommendations for AID clinical trails in [15]

better-informed glucose control decisions [15]. A summary of the main CGM-based KPIs is presented in Table 7.2.

The recommended data sufficiency criteria are at least two weeks of data. Episodes of hypoglycemia and hyperglycemia should be considered using a standard duration of at least 15 min.

Various test scenarios should be considered to evaluate the efficacy of the controller by assessing the impact on TIR values. Regarding the meals, the following scenarios are recommended: (i) sensitivity to the varying size of the meal intake ranging from low to high (15 g to 80 g) carbs values as shown in Fig. 7.5, (ii) error in meal size calculation by considering 75% and 125% of actual carbs intake, and (iii) unannounced meals of varying size should be tested. Additionally, the controller response sensitivity to errors in basal insulin needs to be tested with simulations with 75% and 125% of nominal values. It is important to perform a test considering high initial glucose for assessing the capability of the controller to recover from hyperglycemia back to target without generating a hypoglycemic episode. Similarly, a scenario with low initial glucose needs to be considered as well for assessment.

7.4 Discussion for Future Comparison

Currently, available simulators aim at translating the body of knowledge of the pathophysiology of T1D in a mathematical language for insulin-related and multi-hormone therapy testing. A key element to consider in modeling complex biological/biomedical systems is the trade-off between prediction accuracy and complexity. In addition, the model's ability to describe key phenotypes often relies on the correct model identification using difficult-to-gather clinical data. Among the models in the literature, the Sorensen and UVA/Padova models are the most comprehensive.

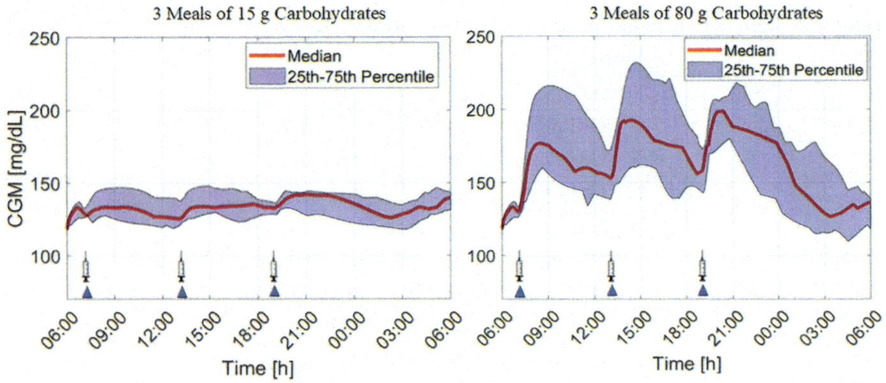

Fig. 7.5 CGM Median and Interquartile Range (IQR) for a simulation including ten adult subjects with varying meal size using T1DMS (S2008c). The meals are given at 07:00, 13:00, and 19:00h, as shown by the triangles. Syringe icons indicate prandial boluses

The latter is, without any doubt, the most used way to test novel therapies in silico aiming at clearance from the regulatory bodies. One outstanding limitation of such models lies in their individualization (identification). Given their complexity, model identification has been conducted by subsystems (rather than considering the whole model) using datasets from diverse studied populations. Moreover, the parameter distributions are unavailable or not in the public domain, which hampers the translation of new technology into clinical practice. Regarding usability, the Hovorka model appears as the simplest, with fewer model parameters, and therefore easier to use for simulation purposes.

Given the difficulty of identifying the parameters of the described maximal models for a particular individual, in silico digital twins of future participants are far to be reached.

AID systems stand at the forefront of diabetes treatment. Commercial systems use a hybrid closed-loop approach, requiring user interaction with the software platform to report meal and exercise events. Nevertheless, ongoing research is directed toward fully autonomous closed-loop systems, aiming to eliminate the need for user intervention, resulting in a burden to the individual. This is driven by maximizing the time spent within the target glucose range, thus reducing the risk of future complications and enhancing the quality of life for individuals with T1D. To do so, the design of the control system plays a critical role.

In summary, AID systems represent the cutting-edge approach for insulin replacement in T1D. Efforts are underway to shift toward fully automated closed-loop systems, with a strong emphasis on addressing the postprandial state, mitigating adverse events to physical activity, and reducing the disease-related burden for the users.

References

1. APS testbed simulator. https://github.com/UVA-DSA/APS_TestBed. Accessed 02 Nov 2023
2. Bergman RN, Ider YZ, Bowden CR, Cobelli C (1979) Quantitative estimation of insulin sensitivity. Am J Physiol Endocrinol Metablsm 236(6):E667
3. Bommer C, Sagalova V, Heesemann E, Manne-Goehler J, Atun R, Bärnighausen T, Davies J, Vollmer S (2018) Global economic burden of diabetes in adults: projections from 2015 to 2030. Diabetes Care 41(5):963–970
4. CGMSIM simulator. https://github.com/lsandini/cgmsim. Accessed 02 Nov 2023
5. Cobelli C, Dalla Man C, Sparacino G, Magni L, De Nicolao G, Kovatchev BP (2009) Diabetes: models, signals, and control. IEEE Rev Biomed Eng 2:54–96
6. Dalla Man C, Rizza RA, Cobelli C (2007) Meal simulation model of the glucose-insulin system. IEEE Trans Biomed Eng 54(10):1740–1749
7. Diabetes Control and Complications Trial Research Group (1993) The effect of intensive treatment of diabetes on the development and progression of long-term complications in insulin-dependent diabetes mellitus. N Engl J Med 329(14):977–986
8. El Fathi A, Smaoui MR, Gingras V, Boulet B, Haidar A (2018) The artificial pancreas and meal control: an overview of postprandial glucose regulation in type 1 diabetes. IEEE Control Syst 38(1):67–85
9. Gregory GA, Robinson TIG, Linklater SE, Wang F, Colagiuri S, de Beaufort C, Donaghue KC, Harding JL, Wander PL, Zhang X, Li X, Karuranga S, Chen H, Sun H, Xie Y, Oram R, Magliano DJ, Zhou Z, Jenkins AJ, Ma RCW, Magliano DJ, Maniam J, Orchard TJ, Rai P, Ogle GD (2022) Global incidence, prevalence, and mortality of type 1 diabetes in 2021 with projection to 2040: a modelling study. Lancet Diabetes & Endocrinol 10(10):741–760
10. Hovorka R, Canonico V, Chassin LJ, Haueter U, Massi-Benedetti M, Federici MO, Pieber TR, Schaller HC, Schaupp L, Vering T, Wilinska ME (2004) Nonlinear model predictive control of glucose concentration in subjects with type 1 diabetes. Physiol Meas 25(4):905
11. Jacobs P (2023) OHSU T1D simulator. https://github.com/petejacobs/T1D_VPP. Accessed 09 Apr 2023
12. Kanderian SS, Weinzimer S, Voskanyan G, Steil GM (2009) Identification of intraday metabolic profiles during closed-loop glucose control in individuals with type 1 diabetes. J Diabetes Sci Technol 3(5):1047–1057
13. Kowalski AJ (2009) Can we really close the loop and how soon? accelerating the availability of an artificial pancreas: a roadmap to better diabetes outcomes. Diabetes Technol & Ther 11(S1):S–113–S–119
14. LoopInsightT1 (LT1) simulator. https://github.com/hpeuscher/loopinsight1. Accessed 02 Nov 2023
15. Maahs DM, Buckingham BA, Castle JR, Cinar A, Damiano ER, Dassau E, DeVries JH, III Doyle FJ, Griffen SC, Haidar A, Heinemann L, Hovorka R, Jones TW, Kollman C, Kovatchev B, Levy BL, Nimri R, O'Neal DN, Philip M, Renard E, Russell SJ, Weinzimer SA, Zisser H, Lum JW (2016) Outcome measures for artificial pancreas clinical trials: a consensus report. Diabetes Care 39(7):1175–1179
16. McGill Artificial pancreas simulator. https://github.com/McGillDiabetesLab/artificial-pancreas-simulator. Accessed 02 Nov 2023
17. Panunzi S, Pompa M, Borri A, Piemonte V, De Gaetano A (2020) A revised Sorensen model: simulating glycemic and insulinemic response to oral and intra-venous glucose load. PLOS ONE 15(8):1–30
18. Patek SD, Lv D, Ortiz EA, Hughes-Karvetski C, Kulkarni S, Zhang Q, Breton MD (2016) Empirical representation of blood glucose variability in a compartmental model. Springer International Publishing, Cham, pp 133–157
19. Perceptus (2021) Glucosym simulator. https://github.com/Perceptus/GlucoSym. Accessed 3 May 2021

20. Resalat N, El Youssef J, Tyler N, Castle J, Jacobs PG (2019) A statistical virtual patient population for the glucoregulatory system in type 1 diabetes with integrated exercise model. PloS one 14(7):e0217301
21. Riddell MC, Peters AL (2023) Exercise in adults with type 1 diabetes mellitus. Nat Rev Endocrinol 19(2):98–111
22. Sorensen JT (1985) A physiologic model of glucose metabolism in man and its use to design and assess improved insulin therapies for diabetes. PhD thesis, Massachusetts Institute of Technology
23. Tyler NS, Jacobs PG (2020) Artificial intelligence in decision support systems for type 1 diabetes. Sensors 20(11)
24. Visentin R, Campos-Náñez E, Schiavon M, Lv D, Vettoretti M, Breton M, Kovatchev BP, Dalla Man C, Cobelli C (2018) The UVA/Padova type 1 diabetes simulator goes from single meal to single day. J Diabetes Sci Technol 12(2):273–281
25. Wilinska ME, Chassin LJ, Acerini CL, Allen JM, Dunger DB, Hovorka R (2010) Simulation environment to evaluate closed-loop insulin delivery systems in type 1 diabetes. J Diabetes Sci Technol 4(1):132–144
26. Woldaregay AZ, Årsand E, Walderhaug S, Albers D, Mamykina L, Botsis T, Hartvigsen G (2019) Data-driven modeling and prediction of blood glucose dynamics: Machine learning applications in type 1 diabetes. Artif Intell Med 98:109–134
27. Xie J (2018) Simglucose v0.2.1. https://github.com/jxx123/simglucose
28. Zhou X, Kouzel M, Ren H, Alemzadeh H (2022) Design and validation of an open-source closed-loop testbed for artificial pancreas systems. In: 2022 IEEE/ACM conference on connected health: applications, systems and engineering technologies (CHASE). IEEE, pp 1–12

Chapter 8
Advances in Run-to-Fail Ball Bearing Testbench: Bridging the Gap in Predictive Maintenance

Javier Diaz-Rozo and Marcela Rondon

8.1 Why Run-to-Fail Testbench?

Ball bearing failure research is critical because they are widely used in many different industries, including automotive, aerospace, and machine-tool industry, among others [1]. In the automotive sector, mechanisms such as engines, transmissions, and suspensions rely on ball bearings, making it crucial to understand and address failures to enhance vehicle performance, efficiency, and safety.

Ball bearings are found in nearly every rotating machine. Thus, the significance of a ball bearing testbench goes well beyond any laboratory setting, holding considerable importance and influence across various industries. Nowadays, one of the industries that is putting more effort in the bearing failure research is the wind power industry, as turbines and generators heavily rely on this component. This sector is facing a big challenge with premature failures on key components that result in an increase in the operating and maintenance costs, which are a priority as they represent up to 30% of the total life cost in both offshore and onshore installations [7]. Also, it has been shown in recent studies that in wind turbines, bearings cause around 70% of gearbox downtime and 20 to 70% of generator downtime depending on its size [8].

Moreover, in the machine-tool industry, a similar scenario unfolds, particularly in grinding machines where the internal ball bearing of the spindle head emerges as a critical component. In a real manufacturing factory, the process of replacing a grinding spindle head can be very time-consuming, taking often many hours. Acquiring a new spindle head may also have a high lead time. It is said that among the components contributing to 26% of grinding machine breakdowns is its spindle head [9].

J. Diaz-Rozo (✉) · M. Rondon
Aingura IIoT, Pº Mikeletegi, 43, 20009 Donostia, Spain
e-mail: jdiaz@ainguraiiot.com

M. Rondon
e-mail: mrondon@ainguraiiot.com

J. M. Maestre and C. Ocampo-Martinez (eds.), *Control Systems Benchmarks*, Advances in Industrial Control, https://doi.org/10.1007/978-3-031-76312-0_8

Regarding these spindle hear failures, faulty bearings are identified as the cause in 42% of the total cases [9]. Thus, by proactively detecting an impending failure, the replacement spindle can be ordered before a defect manifests, enabling planned maintenance and averting disruptions in production.

Similarly, in the aerospace industry, ball bearings play an important role in systems such as aircraft propulsion, making the knowledge extracted from RTF testbenches invaluable to ensure aviation keeps safe and reliable. Also, in the machine-tool industry, where precision and efficiency are critical, the performance of ball bearings significantly affects the overall machining process and thus the efficiency and quality of the manufactured pieces.

Despite the numerous advancements and innovations in the field, ball bearing failure prognostics still pose several challenges due to the complexity of failure modes. Ball bearing failure modes are influenced by various external factors such as air humidity, temperature fluctuations, and external loading. Moreover, there are different ball bearing types, each with its unique physical and geometric specifications like diameter, width, number of rolling elements, material, and type of lubricant [2].

As described before, ball bearings are one of the most basic and most complex mechanisms present in rotating machinery. Furthermore, ball bearings are a single point of failure in all of these mechanisms, meaning that if they fail, the complete machine needs to stop. Therefore, fault detection strategies and solutions to predict ball bearing failures have been developed and published in different publications, such as [3–5]. However, this amount of variability represents a real challenge in order to achieve an appropriate generalization of a ball bearing monitoring solution. This is why researchers in this field are challenged with the task of designing robust and adaptive control strategies for each test, taking into account angular speed, torque, and force, among others, aiming to regulate bearing ball behavior and uncover valuable insights. Therefore, the primary objective to develop a RTF testbench is to research, develop, and validate innovative and effective methods for fault detection in different types of ball bearings.

8.2 Run-To-Fail Testbench Description

Following the aforementioned objective, the RTF testbench outlined in this chapter has a specific design to overcome different challenges described before. The main idea is to facilitate the conduct of accelerated degradation tests on bearings, allowing for testing under both stable and varying operational conditions. Additionally, it needs to accommodate different loads and speed profiles. Figure 8.1 provides a close image of the mechanical system, while Fig. 8.2 shows the overall architecture for the RTF testbench. These subsystems are described in detail in Sect. 8.2.1.

In order to use the data extracted from the testbench, it is fitted with data acquisition, preprocessing, and processing capabilities that are able to

Fig. 8.1 RTF mechanical system

Fig. 8.2 RTF testbench complete view

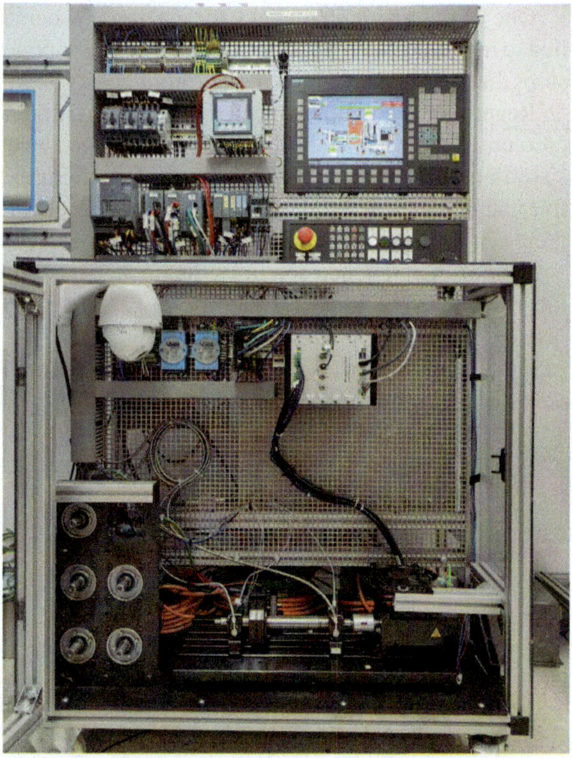

- Acquire data related to ball bearing failures, such as vibration, energy consumption, angular speed, and force, among others. Data must be acquired with appropriate sampling rates, depending on the ball bearing failure frequencies with a minimal jitter.
- Preprocess the data as it arrives from various domains and perform initial sanity checks and clean the data accordingly. Additionally, synchronize the data to ensure that all data and contextual information can be effectively utilized during the processing step. In this step, feature extraction task is done, testing newly designed filtering, transformation, and modeling algorithms.
- Process the features from previous step using the newly developed algorithms that are being tested in the RTF testbench, designed to extract actionable insights, i.e., adding value to data.
- Publish the preprocessed and processed data in order to be able to monitor the test evolution in real time.

These tasks are carried out using a dedicated device designed to manage all of them within the same computing environment, thereby ensuring data integrity and traceability. This device is called Aingura IIoT Edge Computing node, shown in Fig. 8.3, which will be described in Sect. 8.2.5.

One of the major challenges for machine owners is the complexity and expenses tied to maintenance interventions on critical components that can result in unplanned stops that lead to production losses, productivity downturns, and possible safety risks.

Regarding the maintenance of ball bearings, the most common technique is the corrective maintenance, where it is only executed when a failure has already occurred, so by definition it would not help to avoid any failures. The second technique, preventive maintenance, is a time-based broadly used technique with two main weaknesses: there can be unexpected failures due to under-maintenance or rather an unnecessary amount of maintenance activities due to an over-maintenance schedule [6].

Fig. 8.3 Aingura IIoT edge computing node

In that sense, conditioning monitoring is key to detect potential faults by looking for anomalies in the data in early stages to let maintenance personnel take timely actions before an important failure occurs. Conducting accelerated degradation tests in a controlled environment allows for analysis of the system through collected data under various stress conditions, yielding insights that correlate the data with the health state of the machine. This approach can provide actionable insights toward maintenance decision-making, leading to reduced downtime, optimized resource allocation, and, ultimately, higher operational efficiency.

Therefore, the analyzed components of the system are ball bearings. According to SKF,[1] the world's largest manufacturer of this components, ball bearings are used to guide the rotating or oscillating elements of machines, such as shafts, axles, and wheels. They are responsible for transferring loads among the various machine components. A good quality ball bearing must offer high precision and a low friction transmission, enabling elevated rotational speeds while minimizing noise, heat generation, and energy consumption.

The RTF testbench, which can be seen in Fig. 8.2, aims to simulate real-world conditions and investigate the behavior of bearings under various stresses and operational settings. One of the main objectives behind the development of the RTF testbench is to explore innovative and effective ways of fault detection by stressing ball bearings with accelerated degradation tests and collecting data through the process for offline and online analysis. By conducting these tests under stable and varying conditions playing with thermal and mechanical loads, it is possible to gain valuable information about performance limitations and failure modes.

8.2.1 Testbench Mechanical Description

The objective of the mechanical design is to measure the variables (see Sect. 8.2.2) under parameter conditions (see Sect. 8.2.3), in order to obtain information about the remaining useful life. Consequently, it is very important to design a testbench that minimizes result variability while allowing a wide range of test scenarios, yielding high-quality outcomes. Such a testbench demands high attention to ensure that the multiple subsystems function in almost perfect synchronization. These subsystems include the mechanical structure, peripheral equipment, sensors, actuators, and the computing node (see Fig. 8.4). All these components must seamlessly work together to achieve reliable and consistent results.

From the mechanical point of view, the structure of the testbench has six main components: the frame base, the servomotor, the ball bearings, the shaft, the force actuator, and the couplings. A complete view of how all these elements are mounted together can be seen in Fig. 8.5.

The base frame serves as the foundational support for the complete testbench system. It provides stability and ensures the alignment necessary to run the test. It has

[1] https://www.skf.com/.

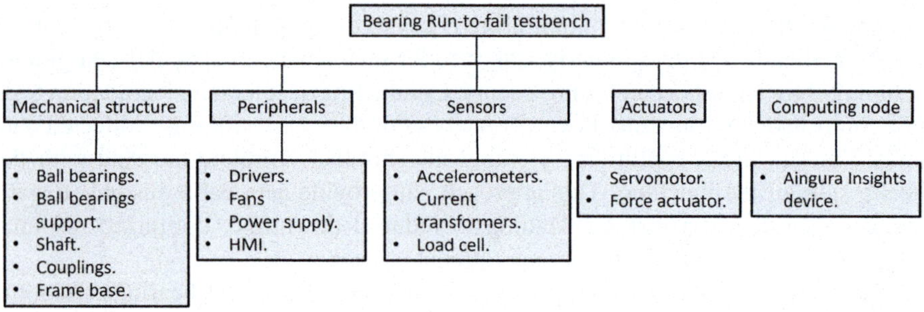

Fig. 8.4 Subsystems comprising the ball bearing failure testbench

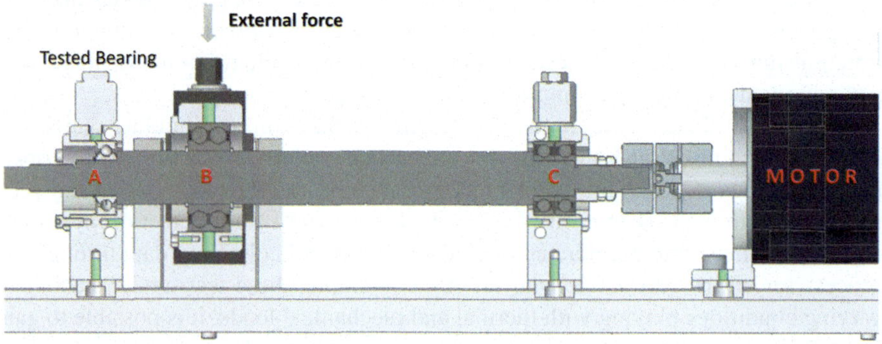

Fig. 8.5 Diagram of the testbench, seen from a frontal view

to be carefully designed to withstand various forces and vibrations generated during the tests to ensure a controlled and repeatable experimental environment. Therefore, it is crucial for the testbench to have optimal rigidity, minimal natural frequencies, and effective isolation from external frequency sources to prevent occurrences like resonance.

The servomotor is the most important part of the system, being responsible for the angular speed and torque applied to the shaft. Thanks to the servomotor drives, a precise long-term control on the angular speed could be maintained. Therefore, the servomotor is essential for creating specific operational conditions in the rotational speeds and to ensure repeatability in experiments. In this testbench, the servomotor is an AC induction motor (see Fig. 8.6).

The ball bearings in position *B* and *C* in Fig. 8.5 are 3206-BD-XL-TVH FAG bearings which are selected for their versatility, low friction, and optimization for minimal noise and vibration, allowing for high rotational speeds. These ball bearings must have the ability to handle both radial and axial loads, with higher useful life and minimum maintenance requirements than the specimen under test (because of its high grade steel, double contact point, and polyamide cage), which is a single row deep grove ball bearing reference 6004, similar to Fig. 8.7.

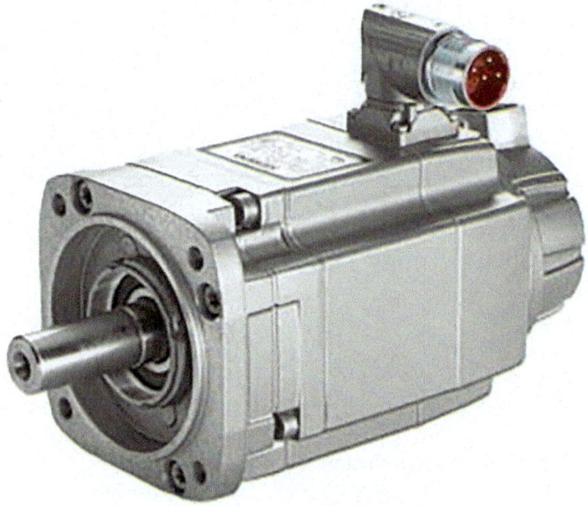

Fig. 8.6 Servomotor Siemens SIMOTICS S-1FK7

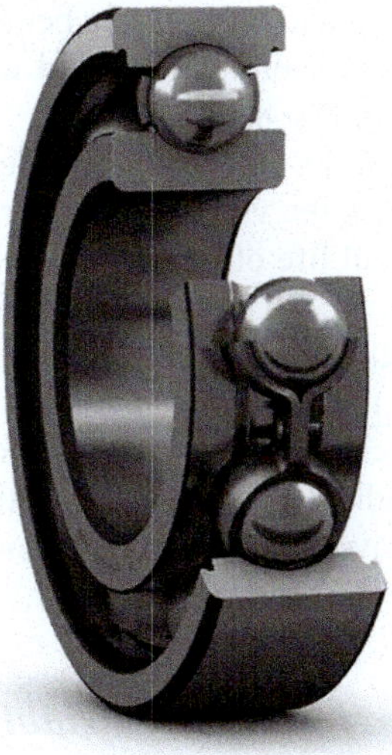

Fig. 8.7 Image of a 6004 ball bearing

The shaft is the central element of the testbench serving as a connector between all mechanical elements. As shown in Fig. 8.5, the shaft has the three different ball bearings mounted in positions A, B, and C. Position A is the run-to-fail specimen location, position B is where the external and radial force is applied by an actuator, and at the right of position C a spider coupling connects it to the servomotor.

Therefore, this shaft is a crucial piece for the proper functioning of the system, which is designed with high precision, considering aspects such as material, diameters, and surface treatments. The material employed is F.1252, a high-strength alloy steel known for its durability. In addition, specific surface treatments, such as

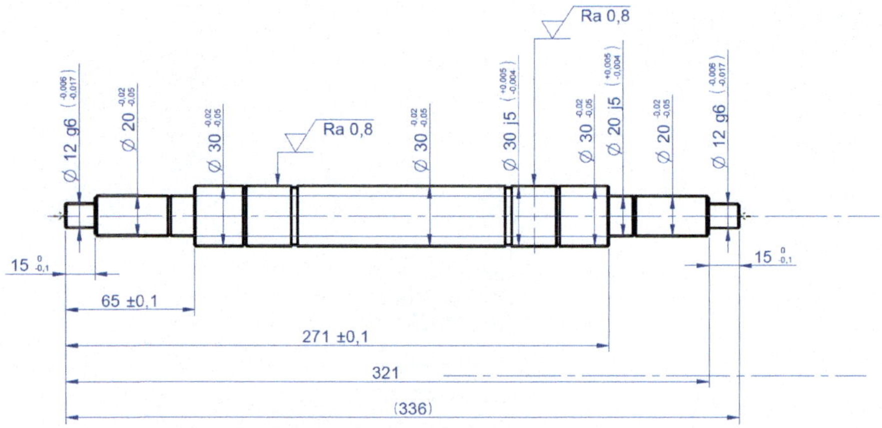

Fig. 8.8 Diagram of the testbench shaft

induction hardening, have been applied to the areas where the bearings are coupled, enhancing their strength and wear resistance. The complete details of the shaft are shown in Fig. 8.8.

The force actuator is the element that can generate load resistance. The actuator facilitates the variation of operational conditions adding axial force that will affect the useful life of the system if it is applied over the ball bearing design values. The actuator is a mechanical element that applies force to the over-sized support bearings to position *B*, as shown in Fig. 8.5.

The couplings are essential for connecting the mechanical components and ensuring efficient transmission of torque. Thus, couplings need to offer substantial angular, longitudinal and radial strength to prevent any clearance between the components and to avoid bending vibrations.

8.2.2 Testbench Variables: Ball Bearing Fault Frequencies

A ball bearing is composed of four main elements: the outer race (1), the balls (2), the cage (3), and the inner race (4), as described in Fig. 8.9. The degradation of these elements results in the emerging of the characteristic failure frequencies in the frequency spectrum. The variation of the amplitudes of these frequencies over time can monitored through the sensors strategically installed in the assembly. Therefore, these amplitudes at specific frequencies are the main variables measured during a test in the testbench.

In order to find each frequency value, the following models are used:

- **Ball Spin Frequency (BSF):** This frequency corresponds to the count of rotations a ball (or roller) completes per shaft revolution

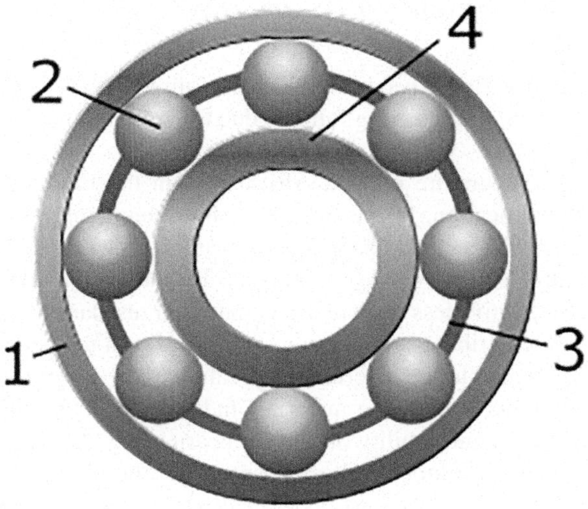

Fig. 8.9 Schematic figure of a ball bearing

$$BSF = \omega \cdot \frac{Pd}{2} \cdot \frac{Bd^2 \cdot \cos^2 2\varphi}{Pd^2},$$

where ω is the angular speed in revolutions per minute (RPM), Pd is the pitch diameter, Bd is the ball diameter, and φ is the contact angle.

- **Ball Pass Frequency Outer (BPFO):** This frequency corresponds to the count of rolling elements passing by a specific point of the outer race during a complete shaft revolution, i.e.,

$$BPFO = \omega \cdot \frac{Nb}{2} \cdot \left(1 - \frac{Bd \cdot \cos\varphi}{Pd}\right),$$

where Nb is the number of balls.

- **Ball Pass Frequency Inner (BPFI):** Similar to the BPFO, the BPFI is the number of rolling elements that passes by a specific point of the inner race, i.e.,

$$BPFI = \omega \cdot \frac{Nb}{2} \cdot \left(1 + \frac{Bd \cdot \cos\varphi}{Pd}\right).$$

- **Fundamental Train Frequency (FTF):** The FTF corresponds to the number of rotations executed by the ball bearing cage per shaft revolution, i.e.,

$$FTF = \omega \cdot \frac{1}{2} \cdot \left(1 - \frac{Bd \cdot \cos\varphi}{Pd}\right).$$

8.2.3 Testbench Parameters

Temperature, speed, and load are three interesting parameters to analyze when studying ball bearing failures due to their significant influence on performance and durability.

Temperature plays a critical role in the ball bearing life as it directly affects its lubrication. In one side, very high temperatures can lead to lubricant evaporation, which can cause increased friction between the ball bearing surfaces. But also, extremely low temperatures can lead to an increase in the lubricant density reducing its effectiveness and creating additional resistance to the movement, which, in turn, will contribute even more to the increase in temperature. It is also worth mentioning that extreme external temperatures may affect both the lubricant condition and the properties of the materials inducing its expansion or contraction, altering the dimensional stability of the ball bearing.

Another important factor, speed, profoundly influences the degradation process of the ball bearing, as high rotational speeds generate elevated forces and stresses in the ball bearing components, intensifying wear and fatigue. High speeds can also contribute to the reduction of the lubrication effectiveness as it may increase the temperature of the grease, allowing for a partial evaporation. As a result, higher speeds will accelerate the fatigue rate in the ball bearing elements.

Additionally, ball bearings are designed to withstand certain loads depending on their application. However, excessive loads can result in heightened stress concentrations, accelerating wear and fatigue in crucial areas. This undesirable stress not only impacts the structural integrity of the ball bearing but also increases the likelihood of surface damage and pitting. That is why here, as we are looking for early failure in an accelerated degradation process, we will surpass the safety thresholds in the parameters previously described.

The RTF testbench has the required capabilities to provide a controlled environment to investigate the effects of the defined operating conditions on ball bearing performance, allowing for precise control over the above-mentioned parameters:

- **Temperature:** The testbench employs an electrical resistance to directly heat the outer surface of the targeted ball bearing, inducing rapid degradation for study.
- **Speed:** The testbench has the flexibility to select a constant rotational speed or program speed variations within the operating ranges of the motor.
- **Load:** The testbench has a manual ball-screw actuator to create a radial force on the ball bearing.

Figure 8.10 shows a layout of the different actuators to produce the required parameters: the heating elements for temperature, the motor, and the external load.

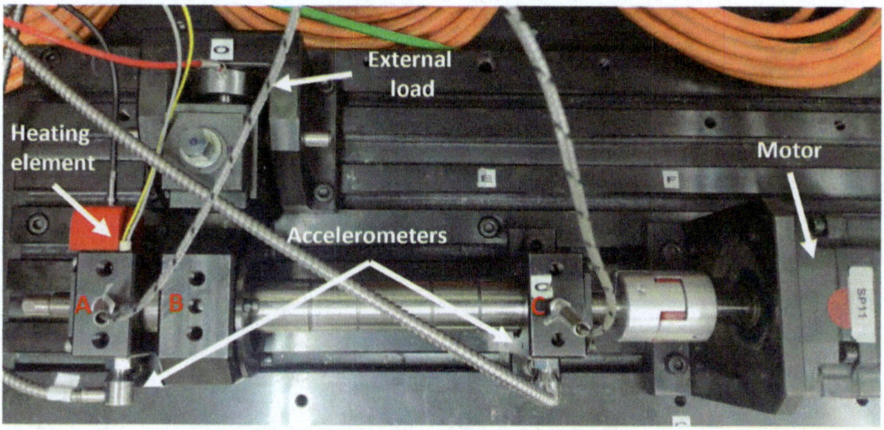

Fig. 8.10 Testbench sensor and actuator layout, with A as the specimen ball bearing, and B and C shaft supports

8.2.4 Testbench Controlled Environment

The system used to control the RTF testbench is called numerical control unit or NCU. In this case, a SIEMENS NCU 840D is used to control different testbench parameters, operation, and security. The integration of the NCU in the complete testbench is described in Fig. 8.11.

The NCU's main function is to interpret the numerical control programs and translate them into precise movements of the axes and other machine functions, such as the control of tools and auxiliary accessories. In this case, the role is to precisely control the speed, described as a parameter in Sect. 8.2.3. In this case, the NCU executes the program instructions while offering an interface to the operator to interact with the machine and supervise its performance. Well known by its design and features, this unit facilitates accurate and efficient control over machine motions and operations, which is essential for modern manufacturing and industrial production. The NCU is integrated by the following elements:

- **Central Processing Unit (CPU):** It is the brain of the NCU and is composed of a robust processor capable of performing complex calculations to control all aspects of the machine operations.
- **Memory:** The NCU has a significant amount of internal storage to save programs, data, configuration parameters, and diagnostic logs.
- **Human–Machine Interface (HMI SINUMERIK):** The NCU has a touch screen or control panel that allows operators to interact with the machine. Through this interface, operators can upload programs, set parameters, monitor machine status, and make adjustments in real time.

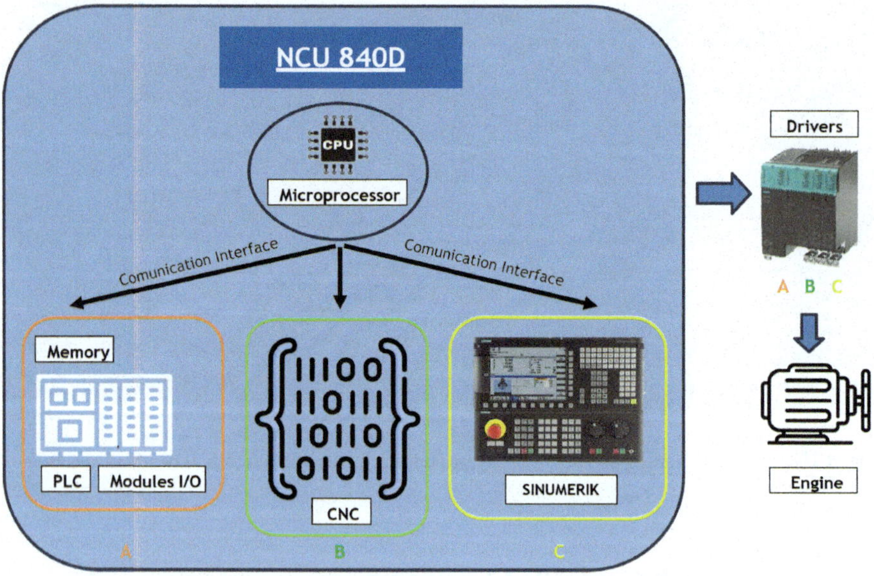

Fig. 8.11 Testbench control flowchart

- **Input/Output Modules:** These modules allow the NCU to communicate with machine peripheral devices such as sensors, encoders, actuators, switches, and other electrical and electronic components.
- **Drivers of Axes (DRIVERS):** The NCU is capable of controlling multiple axes (typically up to 18) on a machine tool. Each axis has its own controller, which manages the movement and precise position for that axis.
- **Communication Interfaces:** It can be equipped with various communication interfaces to exchange data with other interconnected systems or to connect to industrial networks, promoting enhanced interoperability and system integration.
- **Programmable Logic Controller Unit (Integrated PLC):** The NCU has an integrated PLC that allows greater flexibility in terms of automation and coordination with other equipment or processes.

8.2.5 Testbench Edge Computing Node

The RTF testbench includes an Edge-based monitoring device called Aingura Insights (AI). The AI device, shown in Fig. 8.12, is the hardware platform for embedded data collection, analysis, and real-time communication. It is equipped with built-in integrated electronics piezoelectric (IEPE) and energy measurement interfaces, along with several gigabit Ethernet ports for connecting to distributed sensors, PLCs, and industrial networks. A key feature of the AI is its integrated

Fig. 8.12 Aingura Insights (AI) computing node

data storage; a one terabyte internal solid-state drive enables caching any unattended operation in environments with intermittent or null network connectivity. The unit is powered by an external 24 Vdc power supply and can be mounted in an electronics cabinet or portable protective case on a standard DIN rail. The unit measures are 173mm x 125mm x 217mm. This AI computing node has four main capabilities: data acquisition, preprocessing, processing, and actionable insights delivery, which are described below.

8.2.5.1 Data Acquisition

Data acquisition is a critical part of the whole process. The AI is located on the same grid as the testbench, and is connected with current clamps to the energy consumption and with accelerometers to the ball bearings cages to collect vibrations. In addition, the AI has three Ethernet ports that will be used to have connectivity with the NCU for the acquisition of variables such as force, temperature, and speed.

Additionally, this testbench is equipped with a complete sensor system that consists of seven sensors of various types, conscientiously selected to measure the variables of interest. To begin with, there are two accelerometers (see Fig. 8.13a). Model 607A61 from the brand PCB Piezotronics has a sensitivity of 100 mV/g and a measurement range of ± 50 g. This is the most commonly used type of sensor in rotodynamics applications as it measures vibrations, which provide information about the

(a) Accelerometer (b) Thermocouple (c) Current transformer

Fig. 8.13 Testbench sensors

dynamic behavior of the ball bearing. When selecting the accelerometers for the system, factors such as the acceleration range, the frequency range of the vibrations, and external operational conditions must be taken into account for the proper selection of this instrument.

The sensor system includes three thermocouples (see Fig. 8.13b) strategically positioned to measure the temperature magnitude and fluctuations in the outer race of the ball bearing, which is particularly important considering the impact of temperature on the ball bearing lubricants. Also, a strength gauge is also integrated to monitor the force exerted by the actuator on one of the support ball bearings. The presence of this external force helps studying the system performance under diverse stress scenarios. Furthermore, a CTSCMICRO19-250/5-T current transformer (see Fig. 8.13c) is coupled to the system to measure the ball bearing changes from the energy consumption point of view. The data gathered from this sensor can reveal potential system faults by studying the transient behavior of current consumption.

8.2.5.2 Preprocessing

In the signal preprocessing phase, a series of transformations are carried out on the raw vibration signal collected by any of the sensors. The type of preprocessing that is carried out will depend on the type of signal and analysis that will be performed later, as well as what types of insights are sought. The Aingura Insights system has a range of possibilities for signal processing; among them are

- **Vibration-Based Features:** This algorithm used here focuses on monitoring ball bearing fault frequencies based on known operating conditions and vibration data acquisition. To monitor ball bearing fault frequencies, it is necessary to know their geometric characteristics, certain machine operating parameters, and data acquisition. The raw signal is collected by vibration sensors at a sampling frequency of 19531.25 Hz. Upon receiving the raw acceleration signal, it undergoes various processing steps, including modulation, filtering, and the application of the Fourier transform, among others. This is done to obtain a signal where ball bearing fault frequencies are not obscured by noise or ghost frequencies, making the identification of their amplitude less complex.

- **Energy-Based Features:** In signal processing, the power spectrum of a continuous-time signal reveals the distribution of power across its frequency components; by utilizing Fourier analysis, any signal can be decomposed into discrete frequencies or a continuous spectrum [10]. The energy spectral density, calculated when a energy signal is confined within a finite period, provides insight into the distribution of spectral energy per unit time. Thus, the power spectral density (PSD) algorithm indicates the distribution of spectral energy per unit time, as the total energy of such a signal over an infinite period would typically be infinite.
- **Statistical Features:** The root mean square value (RMS) has a precise definition related to the statistical description of periodic signals, especially in electrical signal analysis, and its practical application in vibration analysis positions it as the most widely used numeric value for describing the average vibration level of a machine. The RMS values obtained hold significant utility as they are directly associated with the energy content of the vibration profile, reflecting the destructive potential of vibration, with an additional plus that RMS considers the time history of the waveform.

8.2.5.3 Processing

After the data preprocessing conducted by the Aingura Insights module, the subsequent phase involves data processing. The type of algorithm or method used depends on the type of data that is being processed and what is sought to be obtained after processing. Below will be a brief explanation of some of the algorithms that are currently being used.

- The Gaussian-based Dynamic Probabilistic Clustering (GDPC) algorithm is based on the integration and adaptation of algorithms for time series known for their use in dynamic data analysis scenarios. Among them are the generation of a Gaussian mixture model (GMM) using the Expectation Maximization algorithm to estimate the model parameters and the Page–Hinkley method for anomaly detection. This method can exploit knowledge obtained from engineering processes at different levels and work with data streams, where the EM algorithm is launched only if a concept drift is detected [13].
- The auto-regressive asymmetric linear Gaussian HMM (AR-AsLG-HMM) algorithm employs context-specific linear Gaussian Bayesian networks to factorize emission probabilities. Here, the hidden variable defines the context, and the AR order of each variable for each possible context is determined using the SEM

algorithm and the Yule–Walker equations during score optimization. Additionally, the likelihood function is modified in this model, with adaptations to the forward–backward, Viterbi, and EM algorithms. It has been demonstrated that the AR-AsLG-HMM yields favorable results in health indexes when applied to both synthetic and real data, indicating that the algorithm achieves its intended purpose as a model with a reasonable amount of parameters, offering a robust fit for new data [11].

- The semiparametric Bayesian networks generalize two well-known types of Bayesian networks: Gaussian Bayesian networks and kernel density estimation Bayesian networks. The algorithm accurately learns the combination of parametric and nonparametric components, while achieving a performance comparable with those provided by state-of-the-art methods. Here, Bayesian networks can be used to jointly model uncertain domains with discrete and continuous random variables [12].

8.2.5.4 Actionable Insights Delivery

As shown in Fig. 8.14, the Aingura Insights generates the indicators; all the useful data is sent to a centralized server to be stored and visualized in a helpful way. This server has a MQTT broker running and listening to incoming connections. When the Aingura Insights generates actionable data, it is published to a specific topic in the server. Apart from the MQTT server, there are also other services running that make up the data visualization infrastructure: Telegraf, Influxdb, and Grafana.

- InfluxDB[2] is a time series database that stores all the data received from the Aingura Insights.
- Telegraf[3] is an agent that subscribes to the topics in the MQTT server and writes the received results in the InfluxDB database.
- Grafana[4] is a visualization software that queries the InfluxDB database and draws the data in an easy and understandable way.

8.3 Testbench Use Case

In this section, the operating conditions and results of a real ball bearing accelerated fatigue test conducted on the testbench are shown. Table 8.1 shows the working conditions that were stipulated for this specific test.

The resistance temperature was programmed to remain at 70 °C throughout the test. The shaft's rotation speed was programmed to be a steady 3180 rpm. Notably,

[2] https://www.influxdata.com/.

[3] https://www.influxdata.com/time-series-platform/telegraf/.

[4] https://grafana.com/.

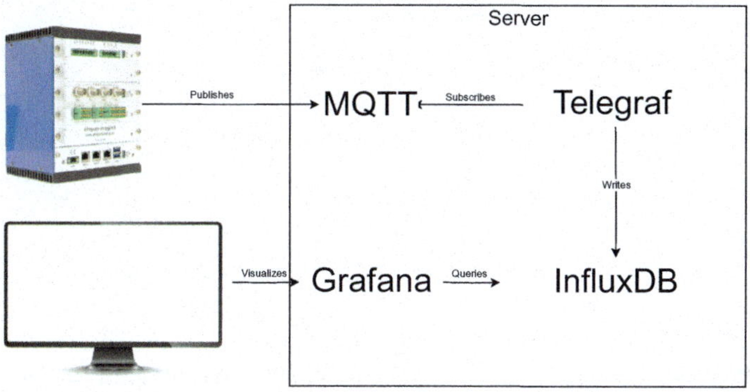

Fig. 8.14 Actionable insights delivery scheme

Table 8.1 Operating conditions for this specific test

Temperature	70 °C
Rotational speed	3180 rpm (53 Hz)
External load	3.5 kN
Lubrication	Normal
Test duration	14 h

the selected speeds for this test are prime numbers, chosen strategically to prevent the rotation speed and its multiples from interfering with other frequencies of interest. Also, the external load exerted on the actuator on the support ball bearing was set at 3.5 kN. The condition of the lubricant was maintained as normal, distinguishing it from other tests where procedures involved extracting or drying the internal grease of the ball bearing to expedite its failure. Finally, with these conditions the test lasted 14 h, which, in the record of test held in this testbench, classifies as short duration. This set of conditions aims to provide a controlled environment for assessing the performance and reliability of the system under specific and well-defined parameters.

As said, the testbench was stopped after 14 h of operation as a consequence of the observed behavior of the collected signals displayed in the online visualization (see Figs. 8.15 and 8.16) and also due to the physical signals such as high noise levels and high engine heating.

The behavior of some of the collected variables such as force, rotation speed, and temperatures can be seen in Fig. 8.15. Also, in Fig. 8.16, the increase in the magnitude of the acceleration RMS can be observed; the trend changes only three hours after starting the test, so it is intuited that that was the instant in which the first crack appeared. Later, that developed until concluding in the ball bearing failures observed in Fig. 8.17, which were taken with a special microscope, after disassembling the damaged ball bearing.

As shown in Fig. 8.17, despite the short duration of the test, the ball bearing elements suffered considerable wear. Most likely, the combination of the three factors,

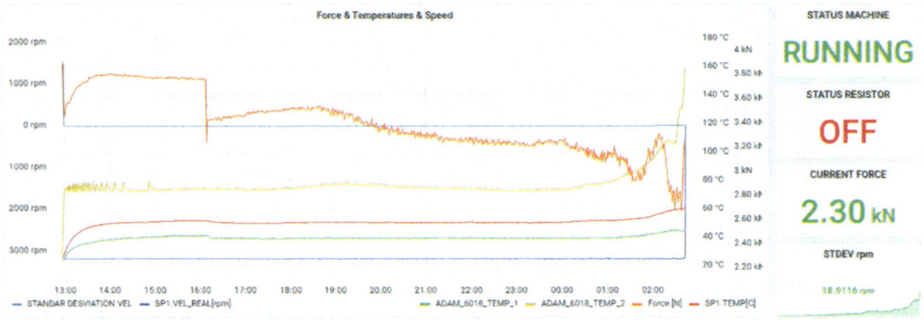

Fig. 8.15 Progress of the temperatures, rotation speed, and load in the last 18 h of the test

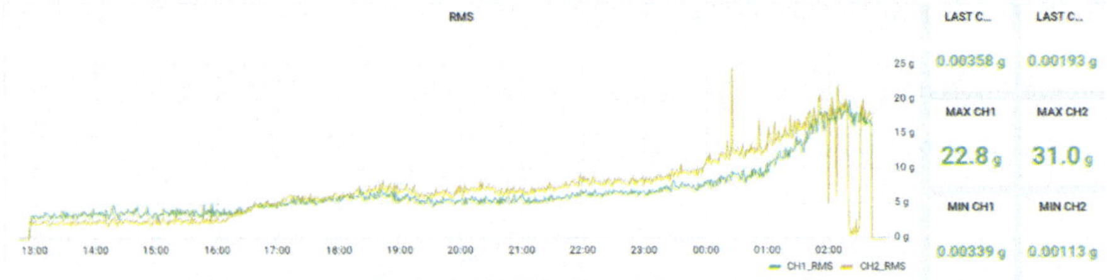

Fig. 8.16 Acceleration RMS for the ball bearing under study

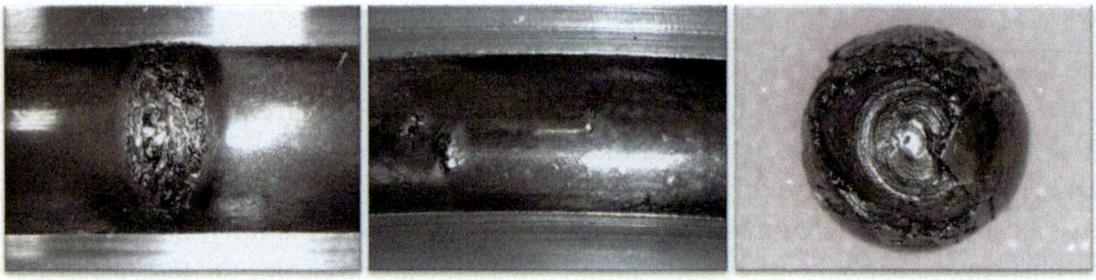

Fig. 8.17 Photographs of different failures in the parts of the ball bearings. From right to left: inner race, outer race, and one of the balls

high temperature, high load, and variable speed, caused the ball bearing to suffer from what is called a high Rolling Contact Fatigue (RCF) very quickly. RCF is one of the predominant modes of failure and is a material removal phenomenon, which is caused by repeated rolling contacts between the rolling elements and the raceways [19]. In this case the most damage was suffer by the inner race and in some of the balls, which lost a great quantity of material.

In Fig. 8.18 a resulting plot of the ball bearing failure frequency amplitude over the course of its operation is shown. There can be observed that during the first three hours they have a consistent behavior, suggesting stability in the early phase of the test. However, a notable shift occurs as the cage frequency (FTF) experiences a sudden decline, which precedes a subsequent increase in all four failure frequencies

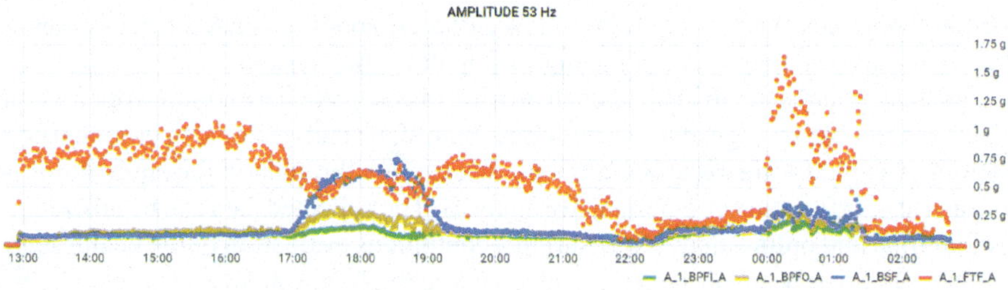

Fig. 8.18 Variation of the four failure frequencies during the test

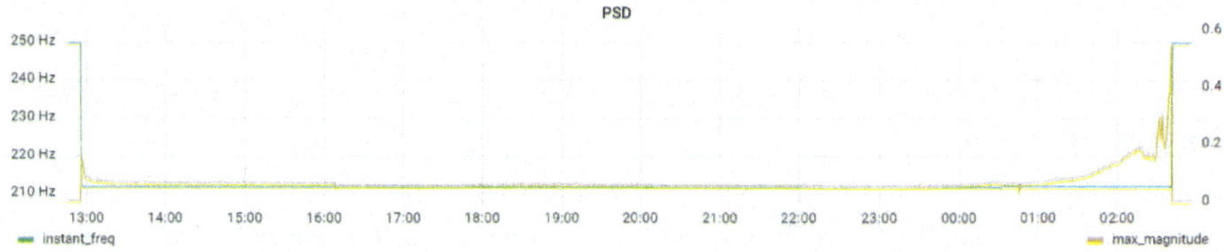

Fig. 8.19 Resulting plot of the PSD algorithm

after an additional hour, which can be an indication of an underlying issue influencing the performance of the ball bearing, such as a developing crack. Subsequently, the four frequencies maintain a relatively constant trend until the eventual failure that occurred 14 h after initiation. At this point, the failure frequencies, particularly the FTF, witness a notable increase, reaching 1.7 g. This kind of visual offers valuable insights into the temporal evolution of the ball bearing performance and highlights critical moments that could signify potential vulnerabilities in the machine.

The obtained amplitudes for the failure frequencies after the application of the power-based algorithm are shown in Fig. 8.19. This is calculated by sliding time windows, with a fixed time window and an overlap percentage. Additionally, for the electrical analysis required in all applications, the power-based algorithm goes through two normalizations, one in amplitude and the other in frequency. Normalization in amplitude is carried out with the amplitude of the fundamental frequency, that is, dividing by the spectral amplitude of the fundamental component of the signal. Frequency normalization is carried out with the fundamental frequency value in Hz. For this reason, the frequency axis of the power-based algorithm will be in terms of the fundamental frequency and its harmonics.

As seen, this kind of test is able to replicate failure trigger such as high temperatures, speeds, and external loads allowing for a complete analysis of the behavior of the ball bearing. By highlighting critical points of interest, like the initiation of cracks and the subsequent ball bearing degradation in the data correlations between operational conditions and ball bearing state can be inferred. The more tests are carried out, the more data of correlations between the variables collected and the state of degradation of the system can be obtained. If there is a high repeatability or a patron between tests with similar operating conditions, it becomes possible to obtain thresholds for some variables that will alert about the state of the system or the monitored element without having to stop the entire system.

As final remarks, the RTF testbench provides a comprehensive framework for conducting ball bearing degradation tests under both static and dynamic operational conditions. The platform enables continuous data capture throughout the entire duration of each test, facilitating comprehensive analysis. The collected data can be utilized for both offline and online analysis, ensuring a continuous evaluation of the health status of the tested ball bearing. This integrated approach offers valuable insights into ball bearing performance and degradation mechanisms, enhancing overall maintenance strategies and reliability assessments.

References

1. Randall RB, Antoni J (2011) Rolling element bearing diagnostics tutorial. Mech Syst Signal Process 25(2):485–520
2. Hamrock BJ, Anderson WJ (1983) Rolling-element bearings (No. NAS 1.61: 1105)
3. Kankar PK, Sharma SC, Harsha SP (2011) Fault diagnosis of ball bearings using machine learning methods. Expert Syst Appl 38(3):1876–1886
4. Sharma A, Amarnath M, Kankar PK (2016) Feature extraction and fault severity classification in ball bearings. J Vib Control 22(1):176–192
5. Kankar PK, Sharma SC, Harsha SP (2011) Fault diagnosis of ball bearings using continuous wavelet transform. Appl Soft Comput 11(2):2300–2312
6. Tian Z, Jin T, Wu B, Ding F (2011) Condition based maintenance optimization for wind power generation systems under continuous monitoring. Renew Energy 36(5):1502–1509
7. de Azevedo HDM, Araujo AM, Bouchonneau N (2016) A review of wind turbine bearing condition monitoring: state of the art and challenges. Renew Sustain Energy Rev 56:368–379
8. Whittle M (2013) Wind turbine generator reliability: an exploration of the root causes of generator bearing failures. Durham theses, Durham University. DurhamE-ThesesOnline: http://etheses.dur.ac.uk/9422/
9. Schwendemann S, Amjad Z, Sikora A (2021) A survey of machine-learning techniques for condition monitoring and predictive maintenance of bearings in grinding machines. Comput Ind 125:103380
10. Broughton SA, Bryan K (2018) Discrete Fourier analysis and wavelets: applications to signal and image processing. John Wiley & Sons
11. Puerto-Santana C, Bielza C, Diaz-Rozo J, Ramirez-Gargallo G, Mantovani F, Virumbrales G, Labarta J, Larrañaga P (2022) Asymmetric HMMs for online ball-bearing health assessments. IEEE Internet Things J 9(20):20160–20177
12. Atienza D, Bielza C, Larrañaga P (2022) Semiparametric bayesian networks. Inf Sci 584:564–582
13. Diaz-Rozo J, Bielza C, Larrañaga P (2020) Machine-tool condition monitoring with Gaussian mixture models-based dynamic probabilistic clustering. Eng Appl Artif Intell 89:103434
14. Jardine AK, Lin D, Banjevic D (2006) A review on machinery diagnostics and prognostics implementing condition-based maintenance. Mech Syst Signal Process 20(7):1483–1510
15. Nectoux P, Gouriveau R, Medjaher K, Ramasso E, Chebel-Morello B, Zerhouni N, Varnier C (2012) PRONOSTIA: an experimental platform for bearings accelerated degradation tests. In: IEEE international conference on prognostics and health management, PHM'12, pp. 1–8. IEEE Catalog Number: CPF12PHM-CDR
16. Gonzalez-Cruz CA, Ceccarelli M, Alimehmeti M, Jauregui-Correa JC (2019) Design and experience of a test-bed for gearboxes. In: Advances in mechanism and machine science:

proceedings of the 15th IFToMM world congress on mechanism and machine science, vol 15. Springer International Publishing, pp. 967–976
17. By Redline—Own work, CC BY-SA 3.0. https://commons.wikimedia.org/w/index.php?curid=1836762
18. Rolling-element bearing (numbered) Niabot, CC BY-SA 3.0. https://creativecommons.org/licenses/by-sa/3.0, via Wikimedia Commons
19. Kang YS (2013) Rolling bearing contact fatigue. In: Wang QJ, Chung YW (eds) Encyclopedia of tribology. Springer, Boston, MA

Part II
Robots and Vehicles

Chapter 9
Fault-Tolerant Control of a Quadrotor Based on a Nonlinear Model

Kenta Hoshino and Toshiyuki Ohtsuka

9.1 Introduction

Fault-tolerant control (FTC) for quadrotors, and multirotors more generally, is an important practical issue for ensuring dependability as these platforms find broader societal applications and roles. A fault-tolerant controller is activated when one or more of the rotors fail, causing deviations in position and attitude, to restore the multirotor's original position and to maintain stable flight. FTC of a multirotor is particularly challenging because a linear model is no longer valid for control design owing to the large attitude motions and limitations on available thrusts. Given the practical and societal demand, this field has been extensively studied for the last decade (see [23, 25] for reviews). The FTC has been addressed through a wide range of approaches, including gain-scheduled PID control [14], observer-based control [9], sliding mode control [12, 22], backstepping-based control [4], and model predictive control (MPC) [3, 10, 17]. Most of the studies in this field address the partial or complete loss of actuators' outputs; the situation is commonly referred to as the loss of effectiveness (LOE) [9, 14]. The gain-scheduled PID control, demonstrated in [14], involves the switching of several PID controllers for quadrotors experiencing LOE. One study [9] presents a nonlinear-observer-based FTC. Similarly, [12] combines a sliding-mode controller with a nonlinear observer to address the FTC under LOE conditions. MPC schemes [3, 10, 17] develop the FTC under LOE, leveraging their inherent flexibility to achieve diverse control objectives. Furthermore, [15] introduces a nonlinear FTC method handling failures in up to three rotors. Considering

K. Hoshino · T. Ohtsuka (✉)
Graduate School of Informatics, Kyoto University, Yoshida-honmachi, Sakyo-ku, Kyoto 606-8501, Japan
e-mail: ohtsuka@i.kyoto-u.ac.jp

K. Hoshino
e-mail: hoshino@i.kyoto-u.ac.jp

J. M. Maestre and C. Ocampo-Martinez (eds.), *Control Systems Benchmarks*, Advances in Industrial Control, https://doi.org/10.1007/978-3-031-76312-0_9

the current trend of the quadrotor FTC, it is necessary to develop a benchmark for comparing the various FTC methods.

In this chapter, we define an FTC benchmark problem for a quadrotor with one rotor failing. The failure of one rotor is a probable accident in practical use of a quadrotor, and its FTC is a baseline for FTC of general multirotors and general failure scenarios. As a benchmark problem, we consider both partial loss and complete loss of a thrust, and we focus on FTC after the rotor failure is detected by some fault diagnosis mechanism. In particular, we deal with a nonlinear model of the quadrotor and also consider the constraints in the thrusts. We first present the nonlinear model and control objectives of the FTC and then define the benchmark problem and performance metrics. We also provide software tools for implementing nonlinear model predictive control (NMPC) as a baseline solution for the benchmark problem. The users can easily modify the software tools to implement and evaluate other control strategies. Finally, we present a design example of the NMPC and assess its efficacy through numerical simulations covering failure scenarios.

9.2 Problem Description

9.2.1 Nonlinear Model of a Quadrotor

The configuration of a quadrotor is depicted in Fig. 9.1. The coordinate axes e^{i}_j and e^{b}_j $(j = 1, 2, 3)$ denote the inertial frame and body-fixed frame, respectively. The state vector of the quadrotor is defined as

$$x = \left[\xi^\top \ v^\top \ q^\top \ \omega^\top\right]^\top \in \mathbb{R}^{13}, \tag{9.1}$$

where $\xi = [\xi_1 \ \xi_2 \ \xi_3]^\top \in \mathbb{R}^3$ and $v = \dot{\xi}$ denote the position and velocity of the quadrotor in the inertial frame, respectively, $q = [q_0 \ q_1 \ q_2 \ q_3]^\top \in \mathbb{R}^4$ denotes the unit quaternion representing the attitude of the quadrotor, and $\omega = [\omega_1 \ \omega_2 \ \omega_3]^\top \in \mathbb{R}^3$ denotes the angular velocity of the quadrotor in the body-fixed frame. The control input vector $u = [u_1 \ u_2 \ u_3 \ u_4]^\top \in \mathbb{R}^4$ consists of thrusts generated by the four rotors. For details of the quaternion representation of the attitude of a rigid body, the readers are referred to [6, 7, 16, 24]. Unlike the Euler angle representation, the quaternion representation has no singularity and is more computationally efficient.

From the Newton–Euler equations of rigid-body motion, the state equation of the quadrotor is obtained as follows [1, 8, 24] (see [13] for a review of the modeling of UAVs):

Fig. 9.1 Configuration of a quadrotor

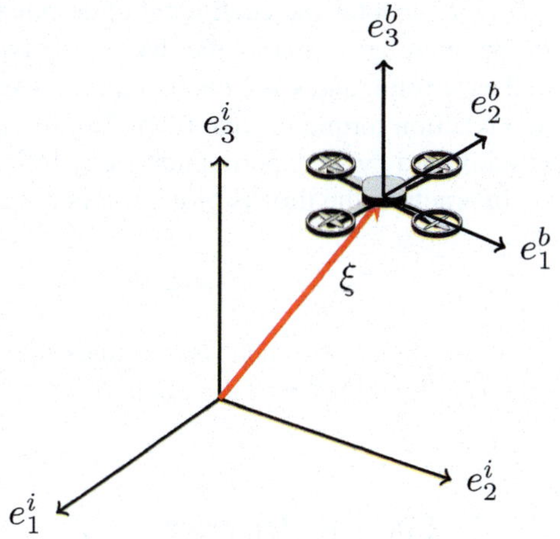

$$\begin{cases} \dot{\xi} = v \\ \dot{v} = -ge_3^{\mathrm{i}} + \frac{1}{m}e_3^{\mathrm{b}}(q)s^\top u \\ \dot{q} = \frac{1}{2}\Omega(\omega)q \\ \dot{\omega} = -J^{-1}\omega^\times J\omega - J^{-1}D\omega + J^{-1}Tu, \end{cases} \tag{9.2}$$

which is also represented in a vector form,

$$\dot{x} = f(x, u), \quad x = [x_1 \ \cdots \ x_{13}]^\top. \tag{9.3}$$

In state equation (9.2), g designates the gravitational acceleration, $e_3^{\mathrm{i}} = [0\ 0\ 1]^\top$ denotes the unit vector of the vertical direction in the inertial frame, m represents the mass of the quadrotor, $J \in R^{3\times3}$ indicates the inertia matrix of the quadrotor, and $e_3^{\mathrm{b}}(q)$, s, $\Omega(\omega)$, $\omega^\times$, D, and T are defined as follows:

$$e_3^{\mathrm{b}}(q) = \left[2(q_0q_2 + q_1q_3) \ \ 2(q_2q_3 - q_0q_1) \ \ q_0^2 - q_1^2 - q_2^2 + q_3^2\right]^\top \in \mathbb{R}^3, \tag{9.4}$$

$$s = [1\ 1\ 1\ 1]^\top \in \mathbb{R}^4, \tag{9.5}$$

$$\omega^\times = \begin{bmatrix} 0 & -\omega_3 & \omega_2 \\ \omega_3 & 0 & -\omega_1 \\ -\omega_2 & \omega_1 & 0 \end{bmatrix} \in \mathbb{R}^{3\times3}, \tag{9.6}$$

$$\Omega = \begin{bmatrix} 0 & -\omega^\top \\ \omega & -\omega^\times \end{bmatrix} \in \mathbb{R}^{4\times4}, \tag{9.7}$$

$$D = \mathrm{diag}[0, 0, d_3] \in \mathbb{R}^{3\times3}, \tag{9.8}$$

$$T = \begin{bmatrix} 0 & \ell & 0 & -\ell \\ -\ell & 0 & \ell & 0 \\ k & -k & k & -k \end{bmatrix} \in \mathbb{R}^{3\times4}, \tag{9.9}$$

where d_3 denotes the coefficient of aerodynamic drag in the yaw axis, ℓ designates the moment arm, that is, the distance between the center of mass of the quadrotor and each rotor, and k is a coefficient representing the relationship between the thrust and reaction torque in each rotor. The use of aerodynamic drag in the present form is supported by an experimental study [15].

In practice, the thrusts of the rotors are constrained within a certain range as

$$u_{\min} \leq u_i \leq u_{\max} \quad (i = 1, 2, 3, 4). \tag{9.10}$$

In most quadrotors, each rotor rotates in one fixed direction and cannot stop completely, which implies that $u_{\min} > 0$.

9.2.2 Control Objective

We consider FTC of the quadrotor to restore and maintain stable flight after some of the rotors loose their thrusts and the position and attitude of the quadrotor deviate from their reference values. In particular, we focus on reduction of one thrust u_1 by a constant thrust reduction factor c_1 $(0 \leq c_1 \leq 1)$ in this chapter. More general cases of failures in the rotors and flight conditions could be considered with suitable modifications of the problem formulation. We assume that the loss of the thrust is identified by some algorithms for fault detection and fault isolation in a certain interval of time after the failure, and the FTC is then activated [25]. We regard the time and state at the activation of the FTC as the initial time $t = 0$ and initial state $x(0) = x_0$ of the FTC problem.

The reduction of thrust u_1 can be represented with vector s and matrix T in state Eq. (9.2) modified to s_{fail} and T_{fail} defined as

$$s_{\text{fail}} = [c_1 \ 1 \ 1 \ 1]^\top \in \mathbb{R}^4, \tag{9.11}$$

$$T_{\text{fail}} = \begin{bmatrix} 0 & \ell & 0 & -\ell \\ -c_1\ell & 0 & \ell & 0 \\ c_1 k & -k & k & -k \end{bmatrix} \in \mathbb{R}^{3\times 4}. \tag{9.12}$$

The terms s_{fail} and T_{fail} coincide with s and T, respectively, when $c_1 = 1$, which implies no reduction in thrust. Additionally, as seen from (9.2), the thrust produced by a rotor specified by u_1 is completely lost when $c_1 = 0$. The constraints in (9.10) are the same for all thrusts $u_1, \ldots, u_4$ before and after the failure, and therefore, u_1 is regarded as a thrust command rather than the actual thrust after the failure.

The control objective of FTC is to regulate the state x from a given initial state $x(0) = x_0$ to a given reference state x_{ref} after the failure, that is, for the state equation (9.2) with s and T replaced by s_{fail} and T_{fail}, respectively. In case of hovering, the reference position ξ_{ref} can be assumed to be zero without loss of generality, and the reference quaternion q_{ref} is given by $q_{\text{ref}} = [\pm 1 \ 0 \ 0 \ 0]^\top$ to maintain the upright

Table 9.1 Physical parameters of the quadrotor

Parameter	Symbol	Value	Unit
Mass	m	0.063	kg
Gravitational acceleration	g	9.81	m/s^2
Moment arm	ℓ	0.0624	m
Reaction coefficient	k	0.0731	m
Moment of inertia	J_1	5.83×10^{-5}	kg·m^2
Moment of inertia	J_2	7.17×10^{-5}	kg·m^2
Moment of inertia	J_3	1.00×10^{-4}	kg·m^2
Drag coefficient	d_3	1.00×10^{-3}	N·m·s
Maximum thrust	$u_{\max}$	0.3266	N
Minimum thrust	$u_{\min}$	0.0065	N

attitude. Note that q_{ref} represents the same upright attitude regardless of the sign of q_0, and, since $\sum_{i=0}^{3} q_i^2 = 1$ holds, it suffices to regulate $q_1 = q_2 = q_3 = 0$ to achieve the upright attitude. Reference velocity v_{ref} and reference angular velocity ω_{ref} are also zero to maintain the reference position and attitude. Therefore, the reference state x_{ref} is given by

$$x_{\text{ref}} = [0\,0\,0\;\;0\,0\,0\;\;\pm 1\,0\,0\,0\;\;0\,0\,0]^\top. \tag{9.13}$$

9.3 Benchmark Design

We employ the Parrot Mambo minidrone[1] as the platform for the benchmark problem. Table 9.1 summarizes its physical parameters taken from a Simulink® Support Package.[2] We assume there is neither disturbance nor modeling error in the system, all state variables are measurable at sampling period of 1 ms without noises, and the thrust reduction factor c_1 is known. We also assume the control inputs $u_1, \ldots, u_4$ are applied without delay at every sampling time.

Benchmark Problem: For a given thrust reduction factor $c_1 \in [0, 1]$, design a feedback controller for the quadrotor to regulate its state from a given initial state $x(0) = x_0$ to the reference state x_{ref}. Evaluate the controller with the following performance metrics for each component of the position, ξ_1, ξ_2, and ξ_3, respectively:

- Steady-state error
- Overshoot

[1] https://www.parrot.com/us/support/documentation/mambo-range.

[2] https://www.mathworks.com/hardware-support/parrot-minidrones.html.

- Settling time
- Undershoot.

When the initial value $\xi_i(0)$ is negative, the performance metrics are defined as follows. The steady-state value $\bar{\xi}_i$ of a coordinate ξ_i is defined by its mean value over a certain time interval before the end of the simulation. For peak value $\xi_i^{\mathrm{peak}} = \max_t \xi_i(t)$ of a coordinate ξ_i, the overshoot is defined by $100(\xi_i^{\mathrm{peak}} - \bar{\xi}_i)/(\bar{\xi}_i - \xi_i(0))$ [%]. The settling time is the minimum time such that $\xi_i(t)$ enters and remains in the range of $\pm 2\%$ from $\bar{\xi}_i$. For the minimum value $\xi_i^{\min} = \min_t \xi_i(t)$ of a coordinate ξ_i, the undershoot is defined by $100(\xi_i(0) - \xi_i^{\min})/(\bar{\xi}_i - \xi_i(0))$ [%]. In particular, the undershoot of ξ_3 indicates the altitude loss at the beginning of the control due to the rotor failure.

Note that the state of the quadrotor may not converge to a constant value due to the unbalanced thrusts and reaction toques generated by the normal rotors. In particular, the quadrotor can keep rotating around the e_3^{b} axis, inducing rotations around other axes. Therefore, we focus on the performance metrics in the position. Moreover, in case the position also keeps oscillating and does not converge to a constant value, the settling threshold of $\pm 2\%$ might be relaxed to, for example, $\pm 5\%$ or the magnitude of steady-state oscillation might be used as an alternative performance metric.

The initial state is given by

$$x_0 = [-1\ -1\ -1\ \ 0\,0\,0\ \ 1\,0\,0\,0\ \ 0\,0\,0]^{\top} \tag{9.14}$$

for the normal operation and FTC. In the case of FTC, a more critical situation can be evaluated, for example, with an initial state,

$$x_0 = \left[-1\ -1\ -1\ \ 0\,0\,0\ \ \cos\frac{\theta_0}{2}\ 0\ \sin\frac{\theta_0}{2}\ 0\ \ 0\ \omega_{20}\ 0\right]^{\top}, \tag{9.15}$$

where θ_0 and ω_{20} denote the initial values of the pitch angle and its angular velocity, respectively.

9.4 Accessing the Benchmark

Jupyter notebooks for simulating the FTC of the quadrotor are provided at a GitHub repository.[3] This repository is based on the automatic code generation tool for NMPC, AutoGenU for Jupyter[4] [11], and contains the following notebooks for the present benchmark problem:

- `QuadrotorFTC.ipynb`
- `RunSimQuadrotorFTC.ipynb`

[3] https://github.com/ohtsukalab/QuadrotorFTC_Benchmark.

[4] https://ohtsukalab.github.io/autogenu-jupyter.

- `PlotSimQuadrotorFTC.ipynb`.

`QuadrotorFTC.ipynb` is a notebook for automatic generation of C++ code and its Python interface for NMPC of the quadrotor as well as the closed-loop simulation by the generated C++ code. A design example and simulation results of NMPC will be given in the next section as the baseline. `RunSimQuadrotorFTC.ipynb` is a notebook for the closed-loop simulation by the Python interface generated by `QuadrotorFTC.ipynb`, and `PlotSimQuadrotorFTC.ipynb` is a notebook for plotting a simulation result by `QuadrotorFTC.ipynb` or `RunSimQuadrotorFTC.ipynb`. Although `RunSimQuadrotorFTC.ipynb` uses the Python interface generated by `QuadrotorFTC.ipynb` for NMPC, it can also be used for the closed-loop simulation of any other user-defined controllers in Python. `PlotSimQuadrotorFTC.ipynb` can also be used for plotting the closed-loop responses of the user-defined controllers.

First, a user needs to set up a JupyterLab or Jupyter Notebook[5] together with a C++ compiler, a Python environment, and some other tools according to the `README.md` in the repository.[6] Then, the user can clone the repository to a local working directory `QuadrotorFTC_Benchmark` with the following command:

```
git clone https://github.com/ohtsukalab/QuadrotorFTC_Benchmark
   -recursive
```

The working directory `QuadrotorFTC_Benchmark` contains the above three notebooks together with various files and directories for automatic code generation.

Next, the user opens `QuadrotorFTC.ipynb` by JupyterLab and runs it to generate C++ code and a Python interface for NMPC. The generated C++ source files are saved in the `QuadrotorFTC_Benchmark/generated/QuadrotorFTC` directory. `QuadrotorFTC.ipynb` also builds the executable file from the generated C++ source files and performs the closed-loop simulation. The log files of the simulation results are saved in `QuadrotorFTC_Benchmark/generated/QuadrotorFTC/log` directory. Then, the user can also run the closed-loop simulation with `RunSimQuadrotorFTC.ipynb`. An advantage of using `RunSimQuadrotorFTC.ipynb` is that the user can change some parameters in the model and NMPC in Python. That is, once a Python interface is generated by `QuadrotorFTC.ipynb`, the user can modify some of the parameters in NMPC without regenerating the C++ source files and Python interface. As the simulation model of the quadrotor is redefined in `RunSimQuadrotorFTC.ipynb`, the user can also perform the closed-loop simulation with model–plant mismatch.

Finally, the user can plot the simulation result in the log files by running `PlotSimQuadrotorFTC.ipynb`. `PlotSimQuadrotorFTC.ipynb` plots time histories of some variables and evaluates the performance metrics in the previous section. It also plots the trajectory of the quadrotor in the three-dimensional space and generates its animation.

[5] https://jupyter.org/.

[6] https://github.com/ohtsukalab/QuadrotorFTC_Benchmark#readme.

For Windows users and Mac users, pre-built Python interfaces are available in the zip files at the release page[7] of the repository. If a user downloads and extracts `QuadrotorFTC_Benchmark_Win.zip` (for Windows) or `QuadrotorFTC_Benchmark_Mac.zip` (for Mac), the user can directly run `RunSimQuadrotorFTC.ipynb` for simulation of FTC by NMPC and `PlotSimQuadrotorFTC.ipynb` for plotting the simulation result. The user does not need to run `QuadrotorFTC.ipynb` for generating C++ code and Python interfaces for NMPC.

9.5 Discussion for Future Comparison

9.5.1 Nonlinear Model Predictive Control

In this section, we employ NMPC as the baseline for FTC of the quadrotor, which can deal with the nonlinear model with input constraints. We define the performance index for NMPC by a simple quadratic function as follows:

$$J = \varphi(x(t+T)) + \int_t^{t+T} L(x(\tau), u(\tau))d\tau, \tag{9.16}$$

$$\varphi(x) = \frac{1}{2}(x - x_{\mathrm{ref}})^\top S_f (x - x_{\mathrm{ref}}), \tag{9.17}$$

$$L(x, u) = \frac{1}{2}(x - x_{\mathrm{ref}})^\top S (x - x_{\mathrm{ref}}) + \frac{1}{2}(u - u_{\mathrm{ref}})^\top R (u - u_{\mathrm{ref}}), \tag{9.18}$$

where S_f, S, and R are weighting matrices, x_{ref} is the reference state defined by (9.13), and $u_{\mathrm{ref}} \in \mathbb{R}^4$ denotes the reference input, respectively. The weighting matrices S_f and S for the state are typically selected to be positive semi-definite, and the weighting matrix R for input is selected to be positive definite. Regarding the input reference u_{ref}, we impose a balance condition

$$c_1 u_{\mathrm{ref}1} + \sum_{i=2}^{4} u_{\mathrm{ref}i} = mg \tag{9.19}$$

so that the sum of the reference thrusts balances the gravitational force. Although reference thrusts satisfying (9.19) are not unique, we adopt the same value for all thrusts as

$$u_{\mathrm{ref}i} = \frac{mg}{c_1 + 3} \quad (i = 1, \ldots, 4). \tag{9.20}$$

[7] https://github.com/ohtsukalab/QuadrotorFTC_Benchmark/releases.

At each sampling time t of NMPC, we solve an optimal control problem (OCP) to find an optimal control input $u_{\text{opt}}(\tau; x(t))$ $(t \leq \tau \leq t + T)$ minimizing the performance index J in (9.16) subject to the state equation (9.3) and the input constraint (9.10). Note that the optimal control $u_{\text{opt}}(\tau; x(t))$ depends on the current state $x(t)$ of the actual system because the initial condition for the OCP to be solved at time t is given by $x(t)$. If we can solve the OCP sufficiently fast within the sampling period, then we give the actual input to the system by the initial value of the optimal control over the horizon $[t, t + T]$ as

$$u(t) = u_{\text{opt}}(t; x(t)). \tag{9.21}$$

Therefore, NMPC is a type of state feedback even if the OCP is solved numerically.

Since it is crucial for implementation of NMPC to solve an OCP in real time within the sampling period, real-time optimization methods tailored for NMPC have been actively studied for more than two decades [18, 19] (see also [2, 20, 21] for various topics in MPC). Our software tools employ the C/GMRES method [18], which is one of the real-time optimization methods for NMPC.

The horizon length T for NMPC is given by a time-dependent function $T = T_f(1 - e^{-\alpha t})$ with $T_f = 0.4$ [s] and $\alpha = 1$, and the parameter for stabilization of the continuation method is given by $\zeta = 1000$ in C/GMRES. The number of discretization steps over the horizon is given by $N = 100$. The inequality constraints on the input u_i are transformed to equality constraints by introducing dummy inputs $u_{\text{d}i}$ $(i = 1, 2, 3, 4)$ as

$$C_i(u_i, u_{\text{d}i}) = (u_i - \bar{u}_i)^2 + u_{\text{d}i}^2 - u_{\text{mm}i}^2 = 0, \tag{9.22}$$

where

$$\bar{u}_i = \frac{u_{\max} + u_{\min}}{2}, \quad u_{\text{mm}i} = \frac{u_{\max} - u_{\min}}{2}. \tag{9.23}$$

A penalty term $-w_{\text{d}} \sum_{i=1}^{4} u_{\text{d}i}$ is also added to the stage cost $L(x, u)$ in (9.18) to avoid a singularity when a dummy input becomes zero. The weights in the cost function (9.16) are selected as

$$S_{\text{f}} = S = \text{diag}(5, 5, 50, 1, 1, 1, 0, 1, 1, 1, 0.1, 0.1, 0.1), \quad R = I, \quad w_d = 100. \tag{9.24}$$

Since the sign of $x_{\text{ref}7} = q_{\text{ref}0}$ is not unique, and it suffices to regulate $q_1 = q_2 = q_3 = 0$, we set the weights for q_0 to be zero.

9.5.2 Simulation Results

Code generation for NMPC and closed-loop simulations were performed on a PC (OS: Windows 11 Pro; CPU: Intel Core i7-1165G7 2.80 GHz; RAM 32 GB). The average computation time per update of NMPC was 0.14 ms for all cases. Therefore, NMPC can be implemented for FTC of the quadrotor in real time for the sampling period of 1 ms.

Figures 9.2, 9.3, and 9.4 show time histories of the position $(x, y, z) = (\xi_1, \xi_2, \xi_3)$, Euler angles (ϕ, θ, ψ) (roll, pitch, and yaw angles), and control inputs in NMPC for normal operation ($c_1 = 1$), partial failure ($c_1 = 0.5$), and complete failure ($c_1 = 0$), respectively, with the same initial state x_0 in (9.14). Here, we use symbol x to denote the coordinate ξ_1 for better readability although it is same as the state vector. Note also that the actual thrusts are given by (c_1u_1, u_2, u_3, u_4). Tables 9.2, 9.3, and 9.4 summarize the performance metrics for each case. In all the cases, NMPC successfully regulates the position of the quadrotor near the reference position

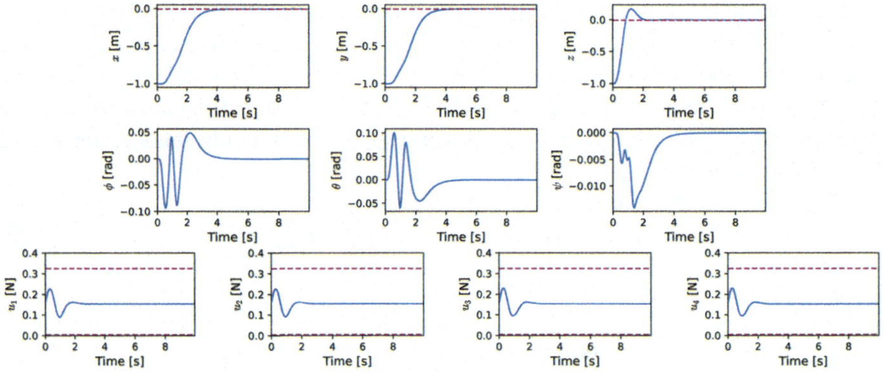

Fig. 9.2 Time histories of nonlinear model predictive control (NMPC) for normal operation, $c_1 = 1$

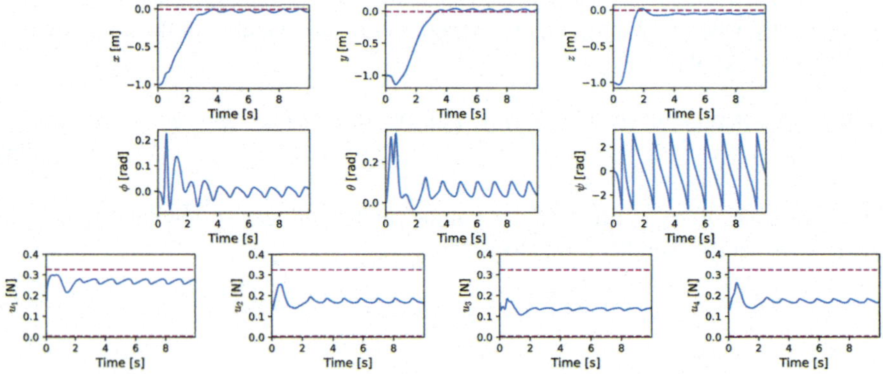

Fig. 9.3 Time histories of NMPC for partial failure, $c_1 = 0.5$

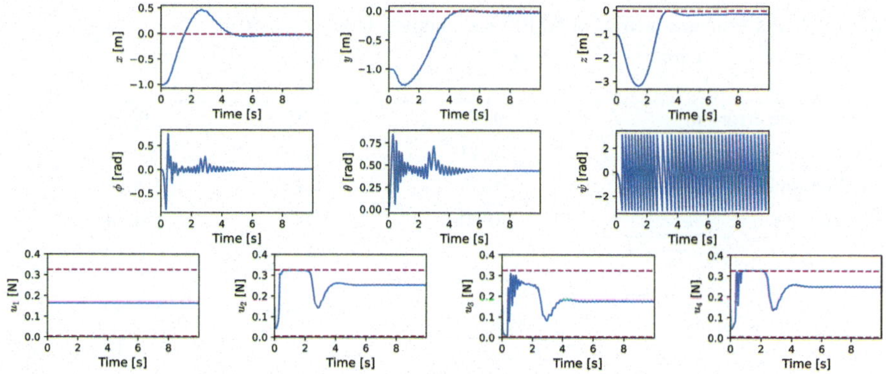

Fig. 9.4 Time histories of NMPC for complete failure, $c_1 = 0$

Table 9.2 Performance metrics of NMPC for normal operation ($c_1 = 1$)

Performance metric	Unit	x	y	z
Steady-state error	m	0.00	0.00	0.00
Overshoot	%	0.00	0.00	17.29
Settling time	s	3.59	3.45	1.92
Undershoot	%	0.00	0.00	0.00

Table 9.3 Performance metrics of NMPC for partial failure ($c_1 = 0.5$)

Performance metric	Unit	x	y	z
Steady-state error	m	−0.02	0.02	−0.05
Overshoot	%	2.59	2.53	7.42
Settling time	s	9.42	4.74	3.54
Undershoot	%	0.00	13.49	2.72

without violating the input constraint. In Fig. 9.4, the two inputs u_2 and u_4 saturate at the maximum value, and the altitude drops more than 2 m before recovering, which demonstrates the critical situation of the failure. In Figs. 9.3 and 9.4 for the cases of failure ($c_1 = 0.5, 1$), the attitude angles keep oscillating, and, in particular, the yaw angle ψ keeps rotating due to the unbalance in the thrusts. Although those oscillations in the attitude also induce small oscillations in the position, the position settles within $\pm 2\%$ of the steady state.

Finally, Fig. 9.5 and Table 9.5 present the time histories and performance metrics, respectively, for complete failure ($c_1 = 0$) with the initial state (9.15) of a non-zero initial pitch angle $\theta_0 = \pi/4$ [rad] and a non-zero initial pitch rate $\omega_{20} = 1$ [rad/s]. This is a more critical situation than that in the previous cases starting from hovering.

Table 9.4 Performance metrics of NMPC for complete failure ($c_1 = 0$)

Performance metric	Unit	x	y	z
Steady-state error	m	−0.02	−0.02	−0.15
Overshoot	%	49.43	4.12	18.22
Settling time	s	6.35	6.16	5.08
Undershoot	%	0.00	27.82	254.74

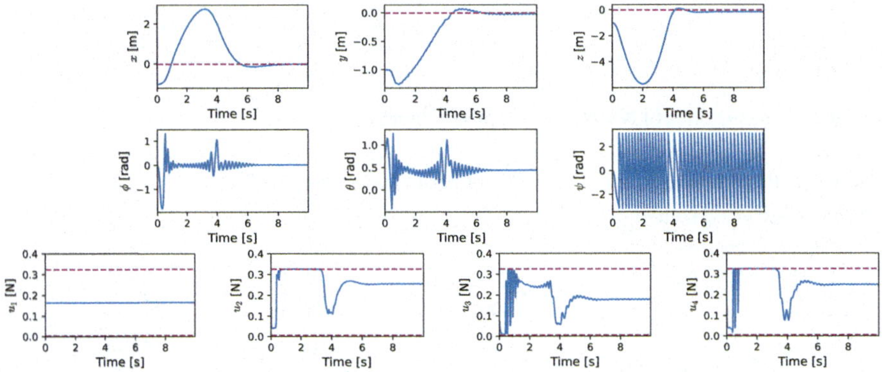

Fig. 9.5 Time histories of NMPC for complete failure, $c_1 = 0$, with $\theta_0 = \pi/4$ [rad] and $\omega_{20} = 1$ [rad/s]

Table 9.5 Performance metrics of NMPC for complete failure ($c_1 = 0$) with $\theta_0 = \pi/4$ [rad] and $\omega_{20} = 1$ [rad/s]

Performance metric	Unit	x	y	z
Steady-state error	m	−0.01	−0.02	−0.15
Overshoot	%	280.09	9.58	2950
Settling time	s	7.94	6.56	6.22
Undershoot	%	0.00	25.70	551.01

Figure 9.5 and Table 9.5 exhibit a significant altitude loss at the beginning of the control and a larger overshoot in x (ξ_1). However, NMPC still regulates the position of the quadrotor successfully near the reference state.

9.5.3 Future Directions

Although NMPC successfully recovers the position of the quadrotor with one rotor failed, there are still some issues to be addressed. First, real-time optimization for

NMPC is computationally demanding and may not be implemented in embedded flight controllers for low-cost quadrotors. Therefore, control methods with low computation costs are preferable as long as they can cope with the nonlinear dynamics and input constraints in quadrotors. In addition to extending traditional nonlinear control methods, it would also be worth exploring appropriate use of machine learning techniques [5]. Second, NMPC in the present setting requires accurate measurements of all the state variables and information about the thrust reduction factor, which are not necessarily available in practice. Integrating state estimation and fault diagnosis and developing feedback control methods based only on available measurements are possible directions of research. Finally, the above-described NMPC has a steady-state error in the position of the quadrotor. Although the steady-state error may be acceptable in the critical situation of a rotor failure, a systematic and simple technique to compensate the steady-state error can be useful in some tasks for a quadrotor.

Acknowledgement This work was partly supported by JSPS KAKENHI, Grant Number 22H01510.

Appendix

This appendix summarizes some properties of the quaternion and Euler angles (see [6, Appendix E] and [7] for details). The coordinate axes of the body-fixed frame are represented in terms of the quaternion $q = [q_0\ q_1\ q_2\ q_3]^\top$ as follows:

$$[e_1^{\mathrm{b}}\ e_2^{\mathrm{b}}\ e_3^{\mathrm{b}}] = \begin{bmatrix} q_0^2+q_1^2-q_2^2-q_3^2 & 2(-q_0q_3+q_1q_2) & 2(q_0q_2+q_1q_3) \\ 2(q_0q_3+q_1q_2) & q_0^2-q_1^2+q_2^2-q_3^2 & 2(-q_0q_1+q_2q_3) \\ 2(-q_0q_2+q_1q_3) & 2(q_0q_1+q_2q_3) & q_0^2-q_1^2-q_2^2+q_3^2 \end{bmatrix}. \tag{9.25}$$

A quaternion $q = [q_0\ q_1\ q_2\ q_3]^\top$ can be converted to ZYX Euler angles consisting of roll angle ϕ, pitch angle θ, and yaw angle ψ as follows:

$$\phi = \arctan\left(\frac{2(q_0q_1+q_2q_3)}{q_0^2-q_1^2-q_2^2+q_3^2}\right), \tag{9.26}$$

$$\theta = \arcsin\left(-2(q_1q_3-q_0q_2)\right), \tag{9.27}$$

$$\psi = \arctan\left(\frac{2(q_1q_2+q_0q_3)}{q_0^2+q_1^2-q_2^2-q_3^2}\right). \tag{9.28}$$

Conversely, ZYX Euler angles ϕ, θ, and ψ are converted to a quaternion q as follows:

$$q = \begin{bmatrix} \cos(\frac{\psi}{2})\cos(\frac{\theta}{2})\cos(\frac{\phi}{2}) + \sin(\frac{\psi}{2})\sin(\frac{\theta}{2})\sin(\frac{\phi}{2}) \\ \cos(\frac{\psi}{2})\cos(\frac{\theta}{2})\sin(\frac{\phi}{2}) - \cos(\frac{\phi}{2})\sin(\frac{\theta}{2})\sin(\frac{\psi}{2}) \\ \cos(\frac{\psi}{2})\sin(\frac{\theta}{2})\cos(\frac{\phi}{2}) + \sin(\frac{\psi}{2})\cos(\frac{\theta}{2})\sin(\frac{\phi}{2}) \\ \sin(\frac{\psi}{2})\cos(\frac{\theta}{2})\cos(\frac{\phi}{2}) - \cos(\frac{\psi}{2})\sin(\frac{\theta}{2})\sin(\frac{\phi}{2}) \end{bmatrix}. \tag{9.29}$$

References

1. Alaimo A, Artale V, Milazzo C, Ricciardello A, Trefiletti L (2013) Mathematical modeling and control of a hexacopter. In: 2013 International Conference on Unmanned Aircraft Systems (ICUAS), pp 1043–1050
2. Allgöwer F, Findeisen F, Nagy ZK (2004) Nonlinear model predictive control: From theory to application. J Chin Inst Chem Eng 35(3):299–316
3. Aoki Y, Asano Y, Honda A, Motooka N, Hoshino K, Ohtsuka T (2021) Nonlinear model predictive control for hexacopter with failed rotors based on quaternions-simulations and hardware experiments-. Mech Eng J 8(5):21–00204
4. Avram RC, Zhang X, Muse J (2017) Nonlinear adaptive fault-tolerant quadrotor altitude and attitude tracking with multiple actuator faults. IEEE Trans Control Syst Technol 26(2):701–707
5. Baskaya E, Bronz M, Delahaye D (2017) Fault detection & diagnosis for small UAVs via machine learning. In: 2017 IEEE/AIAA 36th Digital Avionics Systems Conference (DASC), pp 1–6
6. Choset H, Lynch KM, Hutchinson S, Kantor GA, Burgard W (2005) Principles of robot motion: theory, algorithms, and implementations. MIT press
7. Fresk E, Nikolakopoulos G (2013) Full quaternion based attitude control for a quadrotor. In: 2013 European Control Conference (ECC), pp 3864–3869
8. Hamel T, Mahony R, Lozano R, Ostrowski J (2002) Dynamic modelling and configuration stabilization for an X4-flyer. IFAC Proc Vol 35(1):217–222
9. Hao W, Xian B (2017) Nonlinear adaptive fault-tolerant control for a quadrotor UAV based on immersion and invariance methodology. Nonlinear Dyn 90:2813–2826
10. Izadi HA, Zhang Y, Gordon BW (2011) Fault tolerant model predictive control of quad-rotor helicopters with actuator fault estimation. IFAC Proc Vol 44(1):6343–6348
11. Katayama S, Ohtsuka T (2020) Automatic code generation tool for nonlinear model predictive control with Jupyter. IFAC Proc Vol 53(2):7033–7040
12. Li T, Zhang Y, Gordon BW (2013) Passive and active nonlinear fault-tolerant control of a quadrotor unmanned aerial vehicle based on the sliding mode control technique. Proc Inst Mech Eng Part I J Syst Control Eng 227(1):12–23
13. Mahony R, Kumar V, Corke P (2012) Multirotor aerial vehicles: modeling, estimation, and control of quadrotor. IEEE Robot & Autom Mag 19(3):20–32
14. Milhim A, Zhang Y, Rabbath C-A (2010) Gain scheduling based PID controller for fault tolerant control of quad-rotor UAV. In: AIAA Infotech Aerospace 2010. AIAA, pp 3530
15. Mueller MW, D'Andrea R (2014) Stability and control of a quadrocopter despite the complete loss of one, two, or three propellers. In: 2014 IEEE International Conference on Robotics and Automation (ICRA), pp 45–52
16. Murray RM, Li Z, Sastry SS (1994) A mathematical introduction to robotic manipulation. CRC press
17. Nan F, Sun S, Foehn P, Scaramuzza D (2022) Nonlinear MPC for quadrotor fault-tolerant control. IEEE Robot Autom Lett 7(2):5047–5054
18. Ohtsuka T (2004) A continuation/GMRES method for fast computation of nonlinear receding horizon control. Automatica 40(4):563–574
19. Ohtsuka T, Fujii HA (1997) Real-time optimization algorithm for nonlinear receding-horizon control. Automatica 33(6):1147–1154

20. Raković SV, Leving WS (eds) (2019) Handbook of model predictive control. Birkhäuser
21. Rawlings JB, Mayne DQ, Diehl MM (2020) Model predictive control: theory, computation, and design, 2nd edn. Nob Hill Publishing
22. Sharifi F, Mirzaei M, Gordon BW, Zhang Y (2010) Fault tolerant control of a quadrotor UAV using sliding mode control. In: 2010 Conference on Control and Fault-Tolerant Systems (SysTol), pp 239–244
23. Shraim H, Awada A, Youness R (2018) A survey on quadrotors: configurations, modeling and identification, control, collision avoidance, fault diagnosis and tolerant control. IEEE Aerosp Electron Syst Mag 33(7):14–33
24. Tayebi A, McGilvray S (2006) Attitude stabilization of a VTOL quadrotor aircraft. IEEE Trans Control Syst Technol 14(3):562–571
25. Zhang YM, Chamseddine A, Rabbath CA, Gordon BW, Su C-Y , Rakheja S, Fulford C, Apkarian J, Gosselin P (2013) Development of advanced FDD and FTC techniques with application to an unmanned quadrotor helicopter testbed. J Frankl Inst 350(9):2396–2422

Chapter 10
A Benchmark on Formation Control of Multi-agent Robotic System

Francisco José Mañas-Álvarez, María Guinaldo, and Raquel Dormido

10.1 Introduction

The use of autonomous robot systems that work cooperatively for different robotics-related tasks has been increasing in recent years. In this context, each entity in the system is called an *agent*, and the system is referred to as *multi-agent robotic system* (MARS). Particularly, the deployment of a formation to complete specified tasks safely and efficiently is one of the major interests, and formation control is a prerequisite to this end. For instance, formation applications can be found in sampling, monitoring, or surveillance tasks [1, 2]. In all these tasks, maintaining the formation of the robots plays a crucial role, and the design of distributed control laws that guarantee the achievement and maintenance of such an objective is an active line of research [3].

There are different ways to characterize a formation that lead to different formation control strategies. If we focus on this characterization in terms of the sensing capability and the topology of agents, the natural question is what variables are sensed and what variables are controlled by the multi-agent systems to achieve the formation, and this gives rise to the following categorization [4]:

- Position-based control: Agents can sense their own positions with respect to a global coordinate system and can actively control them to achieve the desired formation so that the agents can move without interacting with each other.

F. J. Mañas-Álvarez (✉) · M. Guinaldo · R. Dormido
Department of Computer Science and Automatic Control of UNED, Madrid, Spain
e-mail: fjmanas@dia.uned.es

M. Guinaldo
e-mail: mguinaldo@dia.uned.es

R. Dormido
e-mail: raquel@dia.uned.es

J. M. Maestre and C. Ocampo-Martinez (eds.), *Control Systems Benchmarks*, Advances in Industrial Control, https://doi.org/10.1007/978-3-031-76312-0_10

- Displacement-based control: Agents can sense and control relative positions to their neighbors to achieve the desired formation. Since this is defined in terms of desired displacements with respect to a global coordinate system, orientation of the global coordinate systems is required.
- Distance-based control: Agents measure the relative positions to their neighbors, but the variables that are controlled are the inter-agent distances. Therefore, the orientation of local coordinate systems is not necessarily aligned with each other. However, the existing control laws are nonlinear, making the analysis more challenging.

Then, according to the previous classification, the formation in a displacement and a distance-based control scheme is achieved if the desired relative positions or inter-distances are reached asymptotically, respectively. It is assumed that these targets that define the formation are given and that the resulting formation is feasible. However, an interesting alternative problem is how they can be generated from high-level specifications, for instance, to achieve an optimal deployment of the formation over an area of interest.

Regarding the topology, the first classifications yield to directed or undirected formations. If the communication between agents is bidirectional, the movement of the formation will depend on the dynamics and the initial conditions of the agents.

Interaction among agents is usually modeled in terms of a graph. Despite the directionality of the links between agents, graph interconnections can have different meanings and this generally depends on the measurement capabilities. For instance, if the agents are equipped with instrumentation that allows the measurement of the variables that need to be sensed, then communication might not be required. However, if these variables cannot be directly accessed by the sensors, then the agents would need to transmit the measurements to other nodes in the network, and then the topology is interpreted as a communication graph [5]. In this case, aspects that concern the communication such as the protocol or the network structure can affect the control performance, and thus, this is an issue that should be handled with care. Furthermore, the frequency of the communication has a direct impact on the control performance but also on the battery life in autonomous devices.

Moreover, when the number of agents in the system is large, it is usually referred to as a swarm. A major challenge in controlling multi-agent swarms is the synthesis of controllers in a scalable manner, i.e., the control design cannot become more complex for a large number of robots. In this regard, a decentralization of the control architecture is always preferred. In the context of swarm robotics, decentralization usually means that the computation is executed onboard. However, in certain devices the computation capacity is limited, and sometimes, the tasks require a central unit, for example, in external indoor positioning systems. Additionally, when the system includes some virtual agents or Digital Twins (DT), then the concern is in the computational resources required by the simulation tools. Therefore, we can assert that scalability is sometimes not trivial, and the study of how much impact the increase in the number of agents has on control is an interesting aspect.

Despite the interest that MARS has generated in both industry and academia for its potential and suitability in real-world scenarios, there is still a lack of adequate tools to validate, evaluate, and compare different approaches comprehensively. Although several benchmarks can be found in the literature on MARS [6, 7], none of them to the authors' knowledge addresses the formation control problem. Therefore, we consider it important for the control engineering research community to design a benchmark to this end that allows evaluation of the different mentioned challenges that arise in MARS formation task.

Robotic Park (RP) [8], a completely heterogeneous, flexible, and easy-to-use indoor platform recently developed to perform MARS experiments in the Department of Computer Sciences and Automatic Control at UNED, is the proposed experimental framework for this benchmark. It supports virtual, real, or hybrid scheme experiences through the two hybrid frameworks available that enable combining physical (and their DT) and virtual agents in mixed reality (MR) experiences. As the DT are indistinguishable from the real agents, their nature is indistinguishable when integrated into ROS 2.

Thus, the goal of this benchmark is twofold: Firstly, the generation of the target formation based on a given set of specifications; and secondly, the design of control algorithms to achieve the formation. The evaluation of the proposed solutions will take into account the complexity of the algorithms, the control performance, the number of communication messages, or the CPU usage.

This chapter is organized as follows: Sect. 10.2 describes the two problems to address in the benchmark, including an overview of the main components and operation modes in RP. The benchmark design, the experimental setup, the proposed test cases, and the performance metrics used to evaluate the formation control system are presented in Sect. 10.3. Section 10.4 provides access resources to the benchmark, including all necessary files and detailing step-by-step how to carry out a virtual validation on RP. Finally, Sect. 10.5 discusses the results and future directions of this work.

10.2 Problem Description

This section describes the problem that the benchmark is designed to address. Section 10.2.1 covers the Robotic Park platform overview. Sections 10.2.2 and 10.2.3 present a detailed description of how the targets of the formation building and the proposed formation control are carried out.

10.2.1 Robotic Park Platform

Robotic Park Components

Robotic Park, the framework for this benchmark, is a flexible experimental platform for indoor MARS that allows combining different types of vehicles, both aerial micro-quadcopters (Crazyflie 2.1 or DJI Tello) and differential mobile robots (Turtlebot3 or Khepera IV). The different software and hardware entities that make up the platform to cooperate and complete a common task are summarized in Fig. 10.1.

RP architecture allows users to work in a virtual environment (system simulation), in a physical environment (robots in a physical environment), or under a hybrid scheme. The virtual environment includes different simulators, based on Gazebo,[1] Webots,[2] and MATLAB®. In the physical environment, different positioning systems are available: a Vicon's MoCap system[3]; the Lighthouse, an indoor optical-based positioning system by Bitcraze,[4] and the Loco Positioning System[5] also by Bitcraze. RP uses a ROS 2 communication middleware that ensures an effective data transmission between the different components of the system despite their heterogeneity. Each component has a node in the ROS 2 domain that can send or request messages to each other using standard types.

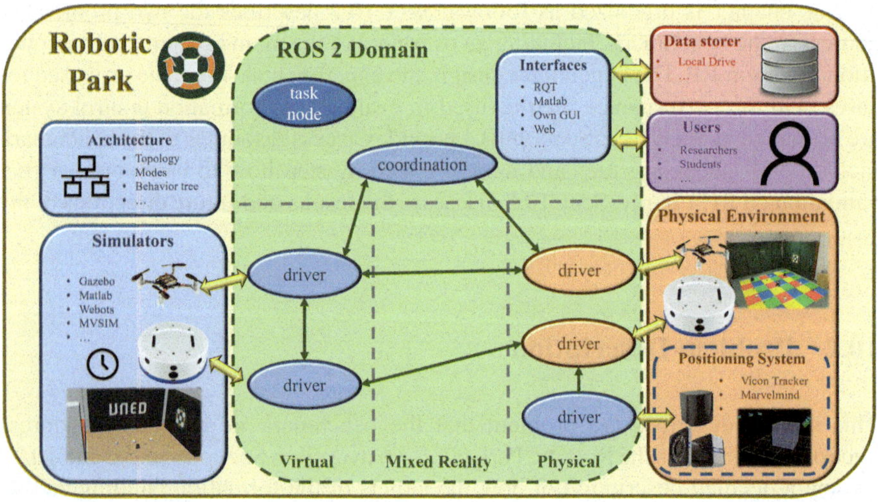

Fig. 10.1 Main components of robotic park

[1] Gazebo: https://gazebosim.org/.
[2] Cyberbotics: https://cyberbotics.com/.
[3] Official Vicon web: https://www.vicon.com/.
[4] Lighthouse: https://www.bitcraze.io/.../positioning/ligthouse-positioning-system/.
[5] LPS: https://www.bitcraze.io/documentation/system/positioning/loco-positioning-system/.

Fig. 10.2 Quadcopter crazyflie

The Crazyflies are the robots used in this benchmark (see Fig. 10.2). They are an open-source nano-quadcopters by Bitcraze suitable for indoor experimentation. Detailed information about their physical characteristics, available sensors, computing, and communication capabilities can be found in [9].

Although we are interested in the design and control of the formation and not in the evaluation of the low-level control of the agents, some references that provide insight into their mathematical model and low-level control of the Crazyflies are available. In this sense, Section 2 of [8] summarizes the two-level control architecture implemented in RP for the Crazyflie. This knowledge will facilitate future expansions of this work.

Robotic Park Operation Modes

The core of the RP architecture is based on the principles of being flexible and modular. The flexibility of the architecture allows for easy and seamless switching between a centralized and distributed topology at both the hardware and software levels. At the hardware level, the user can configure the architecture by selecting the physical computing platform on which each component runs. Modularity in RP implies that changing a component of the system can be done by software without modifying the configuration of the rest of the components.

RP allows three mutually compatible modes of operation: virtual, physical, and hybrid. Their main characteristics are described below.

- **Virtual Mode:** it does not use physical agents, but virtual ones. In this case, the simulators replicate both the components of the physical agents (such as controllers or internal position sensors) and their behavior. The software architecture of the system is designed for the simultaneous execution of the simulator and the rest of the platform's software modules.
- **Physical Mode:** in this mode all agents are physical, which requires specific modules such as positioning systems. In addition, the embedded systems of each agent must be taken into account as extensions of the software structure.

- **Hybrid Mode:** it allows combining virtual and physical agents in the same experience. Using ROS 2 middleware, an agent cannot distinguish whether its neighbors are virtual or physical. The main challenge of this mode of operation is clock synchronization between the two types of agents.

The hybrid mode is particularly interesting in carrying out extensions to the problem addressed in this benchmark. It allows, for instance:

- To scale up the number of agents in the system without the limitation of physical space.
- To test the integration of new robots in combination with those already available.
- To integrate additional virtual sensors into physical robots (cameras, lidars, etc.).
- To teleoperate the real robots in RP remotely with a virtual backup on the user's computer, ensuring stability in case of connection loss.

In this context, RP incorporates DT, virtual agents that can communicate in real-time bidirectionally with physical agents. These twins provide feedback from the virtual sensor to the physical agent, allowing the virtual environment to be perceived locally by the physical agent and represented in the virtual environment.

In RP, defining an experiment requires running a launch file that needs the configuration of a YAML file. This YAML file allows the user to specify the definition of the experience according to their needs. For instance, the operating mode, data logging, CPU monitoring, robots in the experience, etc. can be appropriately defined. A complete definition of this file is presented in the Appendix.

10.2.2 Target Formation Building

We have considered that the target formation for the MARS is a semi-sphere surface since this provides several advantages when flying over an area of interest. A hemispherical surface allows better coverage of the area compared to other geometric shapes, being ideal in applications such as mapping, surveillance, and aerial photography. For instance, it can offer greater safety by providing a constant distance between the object of interest and the drones thus reducing the risk of collisions. It is also especially useful when it is necessary to capture information from different angles and perspectives as clearer visibility of the area of interest from different heights and directions is reached. Also, in applications where efficiency in data capture is important this geometric shape allows a uniform distribution of observation points and minimizes blind spots. Let us model the MARS as a team of N agents denoted by $\mathcal{V}$. The objective of the system is to achieve a formation in the three-dimensional space with the following constraints:

1. All agents should measure the same distance R with respect to a reference point $p_0 = (0\; 0\; z_0)^\top$, $z_0 \geq 0$.
2. In the final configuration, for all agent $i \in \mathcal{V}$, the position $p_i = (p_{i,x}\; p_{i,y}\; p_{i,z})^\top$ should satisfy that $p_{i,z} \geq z_0$.

These two conditions generate a subspace for the deployment of the formation that is the surface of a semi-sphere of radius R and center p_0:

$$x^2 + y^2 + (z - z_0)^2 = R^2. \tag{10.1}$$

Then, the third condition imposed to the formation is the following:

3. The agents should be distributed over the virtual surface of the semi-sphere in such a way that the coverage of the surface approximates the optimal solution.

This requirement connects to a classical optimization problem, the Thomson problem [10]: *find the minimum electrostatic potential energy configuration of N electrons constrained on the surface of the unit sphere*, which has been around for more than a century. This problem seems simple in its formulation, but it is one of the mathematical open problems due to the complexity of the general solution, and the computability or tractability of some simple cases.

If the topology of the MARS is modeled as a static undirected graph $\mathcal{G}$, it is described by the set of agent-nodes $\mathcal{V}$ and the set of edges $\mathcal{E}$. For each agent i, $\mathcal{N}_i$ represents the neighborhood of i, i.e., $\mathcal{N}_i = \{j \in \mathcal{V} : (i, j) \in \mathcal{E}\}$. Note that $|\mathcal{N}_i| = \deg v_i$, where $|\cdot|$ represents the cardinality of the set $\mathcal{N}_i$ and deg is the degree of the vertex v_i associated with the node i.

Then, the proposed solution to this part of the benchmark should generate two sets:

- The set of edges $\mathcal{E}$ for the target formation, which implicitly provides the neighborhood $\mathcal{N}_i$ for each agent i, so that the resulting formation is feasible and ambiguities are minimized.
- The set of target Euclidean distances, denoted as d^*_{ij}, for any pair $(i, j) \in \mathcal{E}$.

So finally, the first problem to solve can be stated as follows:

Problem 1 *Given a MARS whose topology is modeled by a graph $\mathcal{G} = (\mathcal{V}, \mathcal{E})$, find the set of edges $\mathcal{E}$ and the corresponding set of target distances for the formation control, $\{d^*_{ij} : (i, j) \in \mathcal{E}\}$, such that the coverage of the surface* (10.1) *approaches the optimal solution.*

10.2.3 Formation Control

The control architecture considered follows a hierarchical structure, where different control levels are defined, as shown in Fig. 10.3. The controllers defined at the bottom levels are in charge of the positioning of the robots. They are PID controllers implemented on the platform vehicle to track a commanded position. The coordination controller in the upper level receives the pose of the robots and generates a reference for the level below. Since the stabilization controllers are designed and implemented separately, this allows for the consideration of simplified models of the robots for the

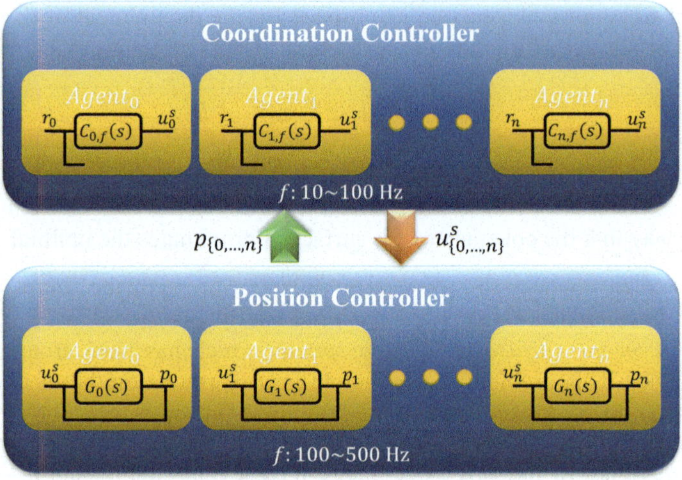

Fig. 10.3 Multi-robot hierarchical control

upper control level. Indeed, single integrator models can be used for the reference positions:

$$\dot{p}_i(t) = u_i(t), \tag{10.2}$$

where $u_i(t) = (u_{i,x}(t)\ u_{i,y}(t)\ u_{i,z}(t))^\top \in \mathbb{R}^3$, i.e., a control signal is generated for each component x, y, and z.

The control objective is defined in terms of a set of target distances, denoted as d^*_{ij}, and generated by the solution to Problem 1. The control problem is solved if each agent i reaches the target distance d^*_{ij}, while the measured distance to the reference point p_0 is R. Thus, the second part of this benchmark can be stated in the following manner:

Problem 2 *Assume that the position controller for each robot of a MARS is given so that the model* (10.2) *can be assumed for the reference positions. Find the distributed control law $u_i(t)$ for the coordination controller and for each agent i*

$$u_i(t) = f_i(p_i, \{p_i - p_j,\ d^*_{ij},\ j \in \mathcal{N}_i\}, p_0, R), \tag{10.3}$$

such that for each neighboring node $j \in \mathcal{N}_i$, the Euclidean distance between them, $d_{ij} = \|p_i - p_j\|$, satisfies

$$\lim_{t\to\infty} d_{ij}(t) = d^*_{ij},\ \ j \in \mathcal{N}_i, \tag{10.4}$$

and also

$$\lim_{t\to\infty} \|p_i(t) - p_0\| = R. \tag{10.5}$$

10.3 Benchmark Design

This section will provide the details for the benchmark for the target formation building and the formation control problems defined in the previous section. For both problems, the number of agents N is set to 15 and the radius of the semi-spherical surface (10.1) is $R = 1$.

10.3.1 Problem 1

The algorithm to solve Problem 1 will be provided in a MATLAB® file (*.m). The input parameters will be $N = 15$ and $R = 1$, and it will generate as outputs the set of edges $\mathcal{E}$ that defines the topology and the set of target distances d^*_{ij}. To evaluate the performance of the proposed solution, the following performance metrics are used:

- CPU time (τ_{CPU}): The time required to execute the proposed solution will provide a good approximation of the complexity of the algorithm. The MATLAB® commands `tic` and `toc` will be used to compute it.
- The distance discrepancy ($\delta(d_{ij})$): It computes the difference between the maximum and the minimum value of the set $\{d^*_{ij}\}$ as a way of measuring the uniformity in the distribution of nodes in the target formation:

$$\delta(d_{ij}) = \max_{(i,j)\in\mathcal{E}} d^*_{ij} - \min_{(i,j)\in\mathcal{E}} d^*_{ij}. \tag{10.6}$$

- Potential energy of the mesh (E_d): As a variation of the Thompson problem [10], only the distance between connected nodes will be considered:

$$E_d = \sum_{(i,j)\in\mathcal{E}} \frac{1}{(d^*_{ij})^2}. \tag{10.7}$$

Thus, the overall cost function J_1 to minimize and, therefore, evaluate the goodness of the proposed solution to Problem 1 is

$$J_1 = \lambda_1 \tau_{CPU} + \lambda_2 \delta(d_{ij}) + \lambda_3 E_d, \tag{10.8}$$

where $\lambda_1 = 2.4283$, $\lambda_2 = 1.0$, and $\lambda_3 = 0.0079$. The parameters were chosen so that each performance index contributes equally to the value of J_1 in the reference case.

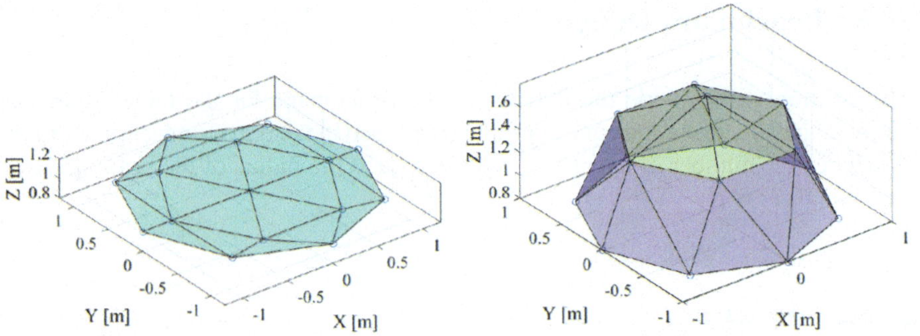

Fig. 10.4 Geometry proposed to the formation benchmark. Agents' initial position (left) and final desired positions (right)

10.3.2 Problem 2

In the formation control problem, we assume that the solution to Problem 1 is received as an input, i.e., the target distances d^*_{ij} and the topology are given.

Initially, the agents are positioned at the height of $z = 0.0$ $[m]$ and are required to first move up to the altitude $z = 1.0$ $[m]$ to reach the stationary coordinates (see Fig. 10.4 left). The agents in the $X - Y$ plane are randomly distributed. The default distribution is available in the experiment configuration file (`config.yaml`) defined in Sect. 10.4. Figure 10.4 shows schematically the proposed initial positions (left) and an example of desired final formation (right), given as a feasible solution to Problem 1.

To evaluate the performance of the proposed control law, the following parameters are considered for each agent i:

- The number of published messages in the network ($n_{m,i}$). Therefore, the defined communication protocol between agents (including the frequency) will be evaluated.
- The number of controller executions ($n_{u,i}$).
- The settling time ($t_{s,i}$), i.e., the time required to reach zero error and remain within 2% of the initial error for each edge. This term is calculated locally in each agent to maintain the distributed architecture of the MARS.
- The Integral Absolute Error (IAE_i). It represents the overall formation error by weighting all errors equally over time. These errors correspond to the deviation of the value of each edge from the desired value. Since the calculation of the errors is done locally, the final result is divided by two to avoid double-counting the error of each edge. It is calculated as follows:

$$IAE_i = \frac{1}{2}\sum_{k=1}^{N_k} \int_0^T \|d^*_{ik} - d_{ik}\| \, dt, \tag{10.9}$$

where k represents the neighbors of agent i and N_k is the number of edges.

- The Integral Time-weighted Absolute Error ($ITAE_i$). This index weights the overall absolute error by the time elapsed since the start of the experience. Errors that exist after a long time are weighted much more heavily than those at the start of the response. This is relevant to our problem as the initial error is always high at the beginning of the formation. Since the error calculation is done locally, the result is divided by two to avoid double-counting the error of each edge. It is calculated as follows:

$$ITAE_i = \frac{1}{2}\sum_{k=1}^{N_k}\int_0^T \|d_{ik}^* - d_{ik}\| t \, dt. \tag{10.10}$$

Therefore, the cost function to evaluate the solutions to Problem 2 is

$$J_2 = \sum_{i=1}^{N}\left(\lambda_1 n_{m,i} + \lambda_2 n_{u,i} + \lambda_3 t_{s,i} + \lambda_4 IAE_i + \lambda_5 ITAE_i\right), \tag{10.11}$$

where $\lambda_1 = 0.02$, $\lambda_2 = 0.1$, $\lambda_3 = 0.3$, $\lambda_4 = 0.1$, and $\lambda_5 = 0.01$. As in (10.8), the parameters were chosen so that each performance index contributes equally to the value of J_2 in the reference case.

The benchmark aims to reduce the cost function (10.11) by designing a coordination control architecture for the proposed MARS. The user should develop an architecture implemented in ROS 2 nodes that communicate with the agents by subscribing to their positions through the topics "*/agent_name/*local_pose" and publishing their control signals as the target positions of each agent through the topics "*/agent_name/*goal_pose".

10.4 Accessing the Benchmark

A public GitHub repository with all the necessary code to access the proposed benchmark can be found in https://github.com/Robotic-Park-Lab/RoboticPark. In order to design experiments in Robotic Park, it is necessary to install ROS 2 Humble on Ubuntu 22.04.

The first step is to install Robotic Park on a local machine. To this end, the repository is cloned and the installation file `install.sh` is executed. This file ensures that all necessary packages are installed, and all of the project is compiled in the workspace. The lines to be executed are the following:

Install Code

```
mkdir -p ~/roboticpark\_ws/src
cd ~/roboticpark\_ws/src
git clone https://github.com/Robotic-Park-Lab/RoboticPark.git
cd RoboticPark
chmod +x install.sh
./install.sh
```

Figure 10.5 shows an overview of the general flowchart architecture to solve Problems 1 and 2 in the proposed benchmark. Different parts can be distinguished. On the left components related to Problem 1 are displayed, and on the right those related to Problem 2. The solution to Problem 1 generates the MARS topology necessary to address Problem 2 as well as the evaluation of Problem 1.

As a second step, it is necessary to define the topology and target distances by solving Problem 1. A MATLAB® function template file `problem1.m` available at the repository mentioned above is provided with two input parameters (`N` and `R`) and three output parameters: (1) `E`, a $N_e \times 2$ matrix where each row denotes an edge (i, j) and N_e is the number of designed edges; (2) `D`, a vector of length N_e where the k-th element is the target distance for the edge in the k-th row of `E`; and (3) `P`, a $N \times 3$ matrix where each row represents the (x, y, z) coordinates of an agent. It provides an example of feasible positioning of the agents over the surface of the semi-sphere that satisfies the configuration defined by `E` and `D`. It is only used for validation of the developed solution to Problem 1.

The third step is to solve the formation control problem defined as Problem 2. The topology, output of Problem 1, is input to the experiment application design in the configuration files of Problem 2. Three configuration files are available (see Fig. 10.5): (1) `World.wbt`, which generates the environment and models of the robots; (2) `Config.yaml`, which allows setting the configurations of the experience (control architecture, controllers, communication protocols, or topology). This file includes the solution to Problem 1; and (3) `Config.rviz` that allows controlling the visualization in the ROS 2 graphical interface RVIZ 2, such as the visualization of 3D models or the settings for the TF visualization. These files allow users to configure any experience on the system. In the middle of Fig. 10.5, the simulation (Webots) and the components, in the form of Nodes within the ROS 2 domain, orchestrating the benchmark are displayed. A ROS 2 launch file written in Python (`experience.launch.py`) is ready to run all the nodes already configured to solve Problem 2 of the benchmark. On the right are the components used to record data, validate data, and evaluate Problem 2. At the bottom, graphs that can be obtained either from the RVIZ 2 interface or from the graphic tool provided by the benchmark itself are shown.

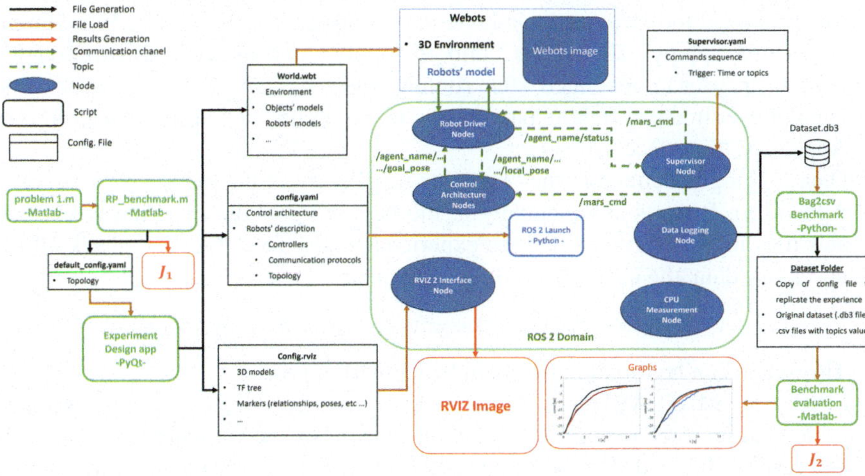

Fig. 10.5 Flowchart of the proposed benchmark in the robotic park for the MARS' formation

A step-by-step summary to run the benchmark is shown below

1. The solution to Problem 1 must be implemented in a MATLAB® function. The script `problem1.m` is provided in the benchmark repository and can be used as a template by the user. Next, you have to run the `RP_benchmark.m` script that evaluates the index (10.8) and generates the default `default_config.yaml` configuration file, for the resulting topology.
2. Modify the configuration file `config.yaml` (see Appendix). The file provided in the base repository is an example configured following the benchmark specifications in the proposed experience. This file defines the MARS topology, providing for each robot in the "Robots" entry the agents with which it is linked and the desired distance from them in the "links" subentry. The user can edit this file to define a new experience by modifying the different entries available: For example, the control nodes, the sampling time, or the links between robots (in the "Robots" entry). Once the file has been modified, it must be located in the "src/RoboticPark/roboticpark config/resources" directory and then compile the workspace so that it can run correctly.
3. Run the startup file with the command

```
ros2 launch roboticpark_config experience.launch.py
```

to execute the ROS 2 launch experience file. If the user does not modify the `config.yaml` provided, the launch file is already prepared to execute Problem 2 of the benchmark with the specifications provided. If the configuration file has been changed by the user, then it is necessary to execute

```
ros2 launch roboticpark_config experience.launch.py
    config_file:='configuration-file'
```

to run the experience.

4. The graphical tool provided available in "roboticpark interfaces/src/Experimental Design app.py" allows the "import" option to load the configuration file (`config.yaml`) generated after running the `RP_benchmark.m` script and manually edit the different parameters to configure the desired display (see Appendix, "Interface" entry).
5. Data logging is defined in the "Data logging" entry of the `config.yaml` (see Appendix). It starts automatically at the beginning of the experience and creates files in .db3 format when the experience ends. To extract and check all the necessary data, the user must run the command

```
python3 bag2csv_benchmark.py dataset_name
```

   This script can be found in the main benchmark repository.
6. Finally, the MATLAB® script `benchmark_evaluation.m` can be executed in the dataset directory for a detailed analysis of the performance index (10.11) and results obtained. This script visually represents the movement of the agents, the formation errors of each agent, the global formation error, and the resource consumption of each registered process.

10.5 Discussion for Future Comparison

There is an increasing interest in swarm systems in different real-world challenges related to rescue missions, surveillance, or search applications. In these scenarios, a MARS instead of a single agent provides benefits to accelerate the goal of the mission or broaden the covered area. Examples of these scenarios can be found in applications for large-scale agriculture in search of weeds or pests [11, 12], border surveillance [13], or wildlife recordings [14]. Most cooperative applications in these problems need the agents to achieve a predefined configuration scheme.

Many publications solve coverage problems for drones to get information about some area of interest. Instead, in this benchmark, we consider Problem 1 of deploying agents over a virtual surface to surveil a particular area of interest. In this sense, the solution to Problem 1 integrates tools that allow the definition of a target formation surface so that the agents are distributed over a semi-sphere. Different approaches have been presented to address the problem of uniform distribution of points [15, 16], but these works do not care about the generation of the topology which is a crucial aspect in this benchmark. Recently, in [17] an algorithm has been proposed for this for a generic quadric surface.

Not only the proper definition of the target formation is important, but the requirement to achieve the formation under optimal agent movement became a challenge. To this end, it is essential to develop and design new control algorithms to achieve the formation efficiently and robustly, both at the individual and collective level. This is the motivation of Problem 2 as it considers the design and evaluation of a formation controller to achieve an optimal deployment over the surface defined in Problem 1.

In this context, several experiences have already been published to study the performance of RP in this type of formation control problem in MARS. Specifically, the scalability, the use of DTs in MR experiences, the control architecture, and the communication protocol have been evaluated [18, 19] that use an adaptation of a distance-based control law [4]. Finally, it is important to highlight that the code and platform provided in this benchmark are ready to accomplish more complex designs and additional research questions for MARS.

As future work, several lines can be considered that allow taking advantage of the flexibility of the Robotic Park platform. Some ideas are the following: Performing experiences in hybrid and physical RP mode, adding heterogeneity to the experiences by using different types of drones available, or evaluating different levels of controller decentralization and different control algorithms. Other extensions of the work have already been commented in Sect. 10.2.1 by using the hybrid mode available in RP.

Acknowledgements This work was supported in part by Agencia Estatal de Investigación (AEI) under the Project PID2020-112658RB-I00/AEI/10.13039/501100011033 and the Project PID2022-139187OB-I00.

Appendix

In RP, starting an experience requires running a global startup file that accepts as an attribute a configuration file written in YAML format. This YAML file contains the following entries to be configured by the user:

- **Operation Mode:** It determines whether the experience is carried out on the real system, in simulation, or as a combination of both. There are three modes available to be chosen by the user: "physical", "virtual", or "hybrid". In the case of the benchmark, the hybrid mode must be chosen.
- **Type of Experience:** There are three available experiences: "navigation", "formation", or "identification". In the case of this benchmark "formation" option must be chosen.
- **Control Architecture:** The control architecture can be defined as centralized or distributed by selecting "centralized" or "distributed_ros". By default, the centralized architecture is selected. In any case, the users can use the centralized node or indicate a set of nodes for distributed structures. In this benchmark the default option is distributed.
- **CPU Monitoring:** Enabling this option allows the user to indicate which processes they want to monitor. These processes must be separated by commas: "process0, process1, process2".
- **Interface:** Two different interfaces are available in RP: "rqt" and "rviz2". Both refer to the standard ROS 2 interfaces and allow loading configuration files to customize the representation of the data.

- **Data Logging:** This option allows the user to define the specific topics they want to record and the name of the recording file. By default, all topics are registered, and the name of the registration is set to the date and time of the trial.
- **Robots:** All robots in the MARS must be defined. Each robot is identified as "robot00", "robot01", etc. Each robot must specify its parameters: name, nature of the robot (physical, virtual, or DT), type of control (with its characteristic parameters), and sampling and the links with the rest of the agents.
- **Supervisor:** It allows the execution of system commands, such as the start of the experience, the takeoff of the aerial robots, and the end of the experience. These commands are implemented in a node that begins the trial when all drones have taken off and reached equilibrium. The test ends 20 s later.
- **Other:** It allows the user to include the execution of their developments in other nodes through the use of entries identified as "node00", "node01", etc.

References

1. Leonard NE, Paley DA, Lekien F, Sepulchre R, Fratantoni DM, Davis RE (2007) Collective motion, sensor networks, and ocean sampling. Proceedings of the IEEE 95(1):48–74
2. Fidan B, Yu C, Anderson BD (2007) Acquiring and maintaining persistence of autonomous multi-vehicle formations. IET Control Theory Appl 1(2):452–460
3. Lawton JRT, Beard RW, Young BJ (2003) A decentralized approach to formation maneuvers. IEEE Trans Robot Autom 19(6):933–941
4. Oh K-K, Park M-C, Ahn H-S (2015) A survey of multi-agent formation control. Automatica 53:424–440
5. Pantelimon G, Tepe K, Carriveau R, Ahmed S (2019) Survey of multi-agent communication strategies for information exchange and mission control of drone deployments. J Intell Robot Syst 95:779–788
6. Yan Z, Fabresse L, Laval J, Bouraqadi N (2015) Metrics for performance benchmarking of multi-robot exploration. In: Proceedings of the RSJ international conference on intelligent robots and systems (IROS). IEEE, Hamburg, Germany, pp 3407–3414
7. Portugal D, Iocchi L, Farinelli A (2018) A ROS-based framework for simulation and benchmarking of multi-robot patrolling algorithms. In: Proceedings robot operating system (ROS) the complete reference, vol 3. Springer, Cham, pp 3–28
8. Mañas-Álvarez FJ, Guinaldo M, Dormido R, Dormido S (2023) Robotic park. Multi-agent platform for teaching control and robotics. IEEE Access 11:34899–34911
9. Giernacki W, Skwierczyński M, Witwicki W, Wroński P, Kozierski P (2017) Crazyflie 2.0 quadrotor as a platform for research and education in robotics and control engineering. In: Proceedings of the 22nd international conference on methods and models in automation and robotics (MMAR). IEEE, Miedzyzdroje, Poland, pp 37–42
10. Thompson JJ (1904) On the structure of the atom: an investigation of the stability and periods of osciletion of a number of corpuscles arranged at equal intervals around the circumference of a circle; with application of the results to the theory atomic Struct Philos Mag Ser 6 7(39):237
11. Albani D, Ijsselmuiden J, Haken R, Trianni V (2017) Monitoring and mapping with robot swarms for agricultural applications. In: Proceedings of the 14th IEEE international conference on advanced video and signal based surveillance (AVSS). Lecce, Italy, pp 1–6
12. Faiçal BS, Pessin G, Geraldo Filho PR, Carvalho AC, Furquim G, Ueyama J (2014) Fine-tuning of UAV control rules for spraying pesticides on crop fields. In: Proceedings of the 26th international conference on tools with artificial intelligence. Limassol, Cyprus, pp 527–533

13. European Commission: Autonomous Swarm of Heterogeneous (2023) https://cordis.europa.eu/project/rcn/209949/factsheet/en
14. Anderson K, Gaston KJ (2013) Lightweight unmanned aerial vehicles will revolutionize spatial ecology. Front Ecol Environ 11(3):138–146
15. Hardin DP, Michaels J, Saff EB (2016) A comparison of popular point configurations on $\mathbb{S}^2$. Dolomites Research Notes on Approximation 9:16–49
16. Koay CG (2011) A simple scheme for generating nearly uniform distribution of antipodally symmetric points on the unit sphere. J Comput Sci 2:377–381
17. Guinaldo M, Sánchez-Moreno J, Zaragoza S, Mañas-Álvarez FJ (2024) Distributed multi-UAV shield formation based on virtual surface constraints. Robot Auton Syst 176:104684
18. Mañas-Álvarez FJ, Guinaldo M, Dormido R, Dormido-Canto S (2023) Scalability of cyber-physical systems with real and virtual robots in ROS 2. Sensors 23(13):6073
19. Mañas-Álvarez FJ, Guinaldo M, Dormido RY, Dormido S (2023) Muestreo y comunicación: impacto en el control de formaciones en sistemas multi-robot heterogéneos. Revista Iberoamericana de Automática e Informática industrial 21(2):125–136

Chapter 11
Digital Twin and Scaled Prototype: Open Low-Cost Benchmarks for Mobile Robotics

Mateo Vernaza and Pablo Velarde

11.1 Introduction

In robotics, the role of control benchmarks is indispensable for evaluating and enhancing the performance of robotic systems. These benchmarks, crucial in design and implementation, offer a platform to gauge performance across various conditions and scenarios. They facilitate a quantitative assessment of control algorithms, enabling the measurement and comparison of key performance indicators like tracking accuracy, stability, response time, efficiency, and robustness. Benchmarks are instrumental in translating theoretical advancements into practical applications by testing these innovations in real and simulated scenarios. They provide a uniform framework for contrasting various control algorithms and systems, catalyzing progress in robotics. More than just a tool for comparison, benchmarks streamline the design process and optimize resources, swiftly pinpointing the strengths, weaknesses, and potential improvements in controllers and algorithms. This underscores the importance of integrating control benchmarks in the field of robotics.

The creation of these benchmarks arose from the necessity for a testing environment for algorithms of a robotic system designed for product collection in a warehouse, developed as part of a thesis project. Given the limited budget, it was crucial to ensure that the control system functioned correctly, as any error could damage the scaled prototype, which was the final testing stage, and severely impact the budget of the project. Consequently, the models described in this chapter were implemented in software to simulate the behavior of a robot. Once the tests on the digital twin

M. Vernaza
Universidad Internacional del Ecuador, Quito, Ecuador
e-mail: mavernazaor@uide.edu.ec

P. Velarde (✉)
Universidad Loyola Andalucía, Seville, Spain
e-mail: pavelarde@uloyola.es

J. M. Maestre and C. Ocampo-Martinez (eds.), *Control Systems Benchmarks*, Advances in Industrial Control, https://doi.org/10.1007/978-3-031-76312-0_11

proved satisfactory, the physical scaled prototype was constructed, leading to the development of the physical version of the benchmark.

Hence, this chapter delves into the two aforementioned benchmarks to establish a comprehensive testing platform for evaluating a robotic system designed for autonomous product retrieval in a warehouse-like environment. The objective was to develop a robot capable of independently navigating an environment, utilizing path planning algorithms and prior knowledge of a predefined *map* detailing the locations of shelves and products. The robot control system was designed to avoid collision with unaccounted obstacles that could be either stationary or dynamically moving through the trajectory, such as people or other robots. In light of the complexity of the requirements, developing a framework to evaluate the system design and performance during and after its implementation became essential.

The first benchmark introduced in this chapter is a scaled, budget-friendly robot prototype, as illustrated in Fig. 11.1 (left). Designed for adaptability, this prototype is well-suited for various tests involving holonomic mobile robots. A central objective was to reduce construction costs through common manufacturing methods while also ensuring the compactness of the prototype. This small size is crucial, allowing for algorithm testing without requiring a full-scale warehouse environment. The robot is equipped with ultrasonic sensors, rotary encoders, and mecanum wheels to align its capabilities with those of the digital twin, facilitating holonomic movement. As a cost-effective option, this prototype significantly enhances testing capabilities and provides an efficient platform for evaluating control algorithms for mobile robots in real-world scenarios.

The second benchmark features a digital twin of the robot, depicted in Fig. 11.1 (right). This digital simulation environment enables swift iterations in the design, fine-tuning, and testing of the control system. The program, implemented in MATLAB® represents the robot and its surroundings in a two-dimensional space.

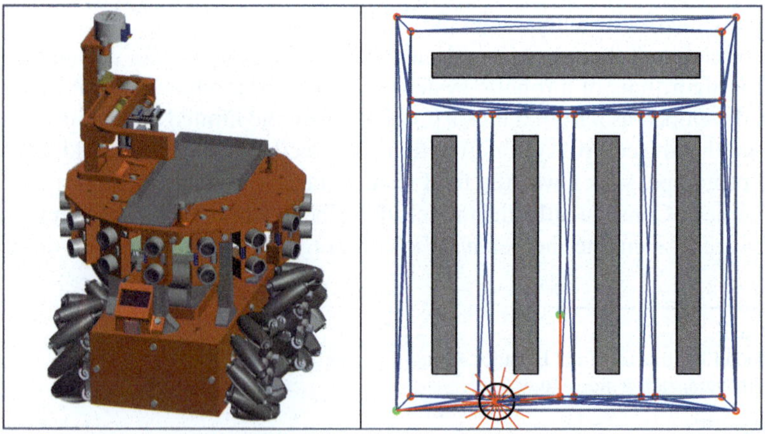

Fig. 11.1 Illustration of the two benchmarking tools: scaled prototype of a holonomic robot (left), digital twin based on MATLAB® (right)

This approach facilitates the creation of varied warehouse-like scenarios with different layouts, product quantities, shelf arrangements, and spatial distributions, allowing for a thorough evaluation of the resilience and efficiency of the robot. The simulation also serves as an open platform to assess control algorithms, particularly those focused on trajectory planning and obstacle evasion, by mimicking the data acquisition of the actual robot. The digital twin simulates two types of sensors: ultrasonic sensors distributed around the robot and rotary encoders that measure wheel rotations, which are utilized to estimate the position of the robot.

Although the physical benchmark is located at Universidad Internacional del Ecuador (UIDE) for students use, both benchmark designs are open-source, allowing anyone to build or modify them as needed. This facilitates testing and learning about robotics and control algorithms. Potential use cases include, but are not limited to, robot position estimation, optimal trajectory planning, trajectory tracking with PID controllers, and obstacle avoidance.

Similarly, recognizing the importance of benchmarks in robotics, various initiatives have been undertaken to establish quantitative performance metrics for mobile robots, as exemplified by the work in [8]. Efforts have also been made to develop simulated benchmarks for robotics, which are crucial for safely simulating autonomous mobile robots. A notable example is the MoVeMA benchmarks introduced by [4]. Additionally, the ROBEL platform presented in [1] offers cost-effective robotic benchmarks specifically designed for testing real-world reinforcement learning applications. These examples highlight significant endeavors contributing to the establishment of benchmarks in the realm of robotics.

11.2 System Description

The digital twin and the scaled prototype are centered around a four-wheeled holonomic robot equipped with mecanum wheels to facilitate omnidirectional movement. This mobile platform incorporates rotational sensors in each wheel motor for accurate measurement of rotational displacement and distance sensors to identify nearby obstacles. It is designed for functionality in standard warehouse environments, typically characterized by even terrain, non-slippery floors, and minimal interference from sensory disturbances such as fog, dust, and intense sunlight.

The benchmarks primarily concentrate on the position control aspect of the robotic system, intentionally setting aside the intricacies of the grasping mechanism. This means the benchmarks are based on the premise that proper robot positioning during retrieval is sufficient to ensure efficient product collection. It is important to note, though, that the challenge of grasping in robotics is far from trivial. This complexity stems from the diverse range of object shapes, sizes, textures, and stiffness. However, to maintain focus, this study limits itself to position control. As such, the simulations do not include the grasping mechanism, and the physical prototype uses a basic gantry mechanism for picking up uniformly shaped objects from shelves.

Given the defined sensing and environmental constraints, the benchmarks focus on three specific areas: trajectory generation, kinematic control of the platform, and, optionally, the avoidance of dynamic obstacles. Essentially, these benchmarks are designed to assess several key aspects: the proficiency of the controller in finding a route to cover all necessary positions within the environment, its ability to adhere to this route accurately, and, if relevant, the capability of the local planner to circumvent unexpected dynamic obstacles while in operation. To address these three key aspects effectively, it is crucial to have a model of the robotic platform and a model of the environment. Therefore, the approach includes developing a kinematic model of the robotic platform for understanding the movement of the robot, an odometry model for estimating the configuration of the robot, and a graph-based environment description for efficient trajectory generation.

11.2.1 Kinematic Model of a Mecanum Wheeled Robot

The kinematic model of a Mecanum-wheeled robot is essential for understanding and controlling its movement. This model allows for the translation of individual wheel speeds into the overall velocity and positional changes. Understanding this relationship between wheel speeds and spatial movement is crucial for effectively managing the motion of the robot.

Defining a method for expressing the state of the robot at any given moment is crucial in this context. This is referred to as the configuration of the robot, which comprises a set of parameters that define the necessary information to understand the state of the robot at an arbitrary time t. For a bi-dimensional mobile robot, this can be represented by the vector $\mathbf{q}(t) = (\phi(t), x(t), y(t))$, where the parameter ϕ represents the orientation robot, and x and y represent its position in a two-dimensional plane at time t. However, it is essential to specify relative to what these parameters are defined. This leads to the concept of a reference frame, a set of coordinates that assists in defining these parameters. In our case, it is crucial to establish a fixed space frame, denoted as $\{s\}$ and depicted in Fig. 11.2, serving as a reference point in the environment (the warehouse). This frame forms the basis for defining the global positions of everything else, including the shelves, the global position of the robot, and the positions of products. Another useful reference frame is the body frame, denoted as $\{b\}$ and also depicted in Fig. 11.2, which is always attached to the chassis position and orientation of the robot. It is particularly useful for understanding small incremental changes in the configuration of the robot, as we will see later. Then, the goal is to find how the movement of each wheel, characterized by its individual speeds, affects the change in the configuration $\mathbf{q} = (\phi, x, y)$ of the chassis relative to $\{s\}$ or $\{b\}$. To accomplish this, a series of transformations of reference frames are performed to establish the wanted relation.

In these transformations, the time derivative of the vector $\mathbf{q}$ is considered, denoted as $\dot{\mathbf{q}} = (\dot{\phi}, \dot{x}, \dot{y})$. This approach aims to relate the vector of the velocities of the wheels

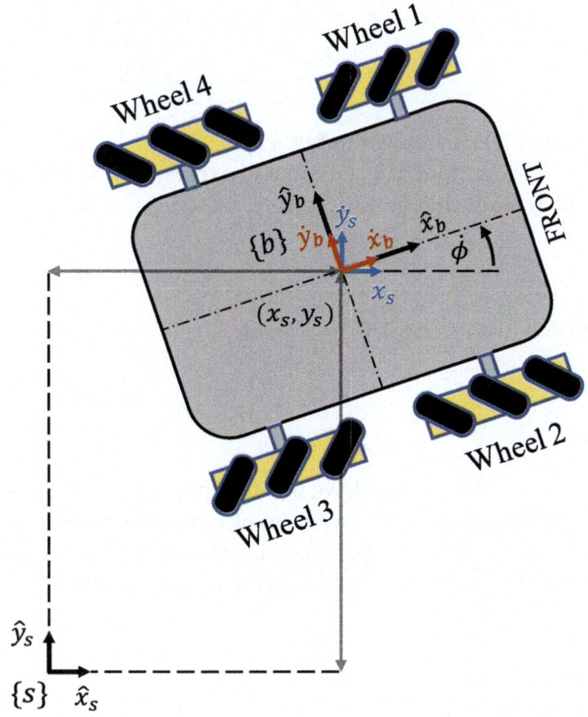

Fig. 11.2 Illustration of the configuration of the robot relative to the space frame $\{s\}$ and the body frame $\{b\}$. The diagram shows the parameters that define the configuration $\mathbf{q} = (\phi, x, y)$ and the relation between the space and body frame

to define the change in the configuration in terms of these speeds. Further details on this will be discussed later.

The first transformation involves a simple rotation operation to express $\dot{\mathbf{q}}$ in $\{s\}$ as a configuration relative to $\{b\}$. This transformation is represented by

$$\dot{\mathbf{q}}^b = \mathrm{R}(\phi) \cdot \dot{\mathbf{q}}, \tag{11.1}$$

where $R_s^b(\phi)$ is the rotation matrix defined as

$$R_s^b(\phi) = \begin{bmatrix} 1 & 0 & 0 \\ 0 & \cos(\phi) & \sin(\phi) \\ 0 & -\sin(\phi) & \cos(\phi) \end{bmatrix}, \tag{11.2}$$

and depicted in Fig. 11.2.

Next, transformations are performed to transition from the $\{b\}$ frame to four different reference frames $\{c_i\}$, each fixed to one of the wheels. These transformations enable the computation of the individual velocities $\dot{\mathbf{q}}^{c_i} = (\dot{x}_{c_i}, \dot{y}_{c_i})$ in each $\{c_i\}$ frame, based on the chassis speeds $\dot{\mathbf{q}}^b$. It is important to note that in these transformations, besides the velocities introduced by $(\dot{x}_b, \dot{y}_b)$, additional tangential velocities arise due to the overall angular velocity $\dot{\phi}_b$ of the chassis. Therefore, the linear velocities in $\{c_i\}$ are

$$\dot{x}_{c_i} = -y_i \cdot \dot{\phi} + \dot{x} = -\dot{x}_{T_i} + \dot{x}, \tag{11.3}$$

$$\dot{y}_{c_i} = x_i \cdot \dot{\phi} + \dot{y} = \dot{y}_{T_i} + \dot{y}, \tag{11.4}$$

and (11.4), as shown in Fig. 11.3.

Furthermore, it is possible to express (11.3) and (11.4) in matrix form, allowing the relationship to be described as

$$\dot{\mathbf{q}}^{c_i} = M_b^{c_i} \cdot \dot{\mathbf{q}}^b, \tag{11.5}$$

where the matrix $M_b^{c_i}$ is given as

$$M_b^{c_i} = \begin{bmatrix} -y_i & 1 & 0 \\ x_i & 0 & 1 \end{bmatrix}. \tag{11.6}$$

Now, it is time to delve into the specifics of Mecanum wheels. These eye-catching wheels are modified versions of common wheels. They translate rotational speed into a linear speed perpendicular to the rotational axis and parallel to the contact surface, as common wheels do. This speed is denoted as $\mathbf{v_{motor}}$. However, Mecanum

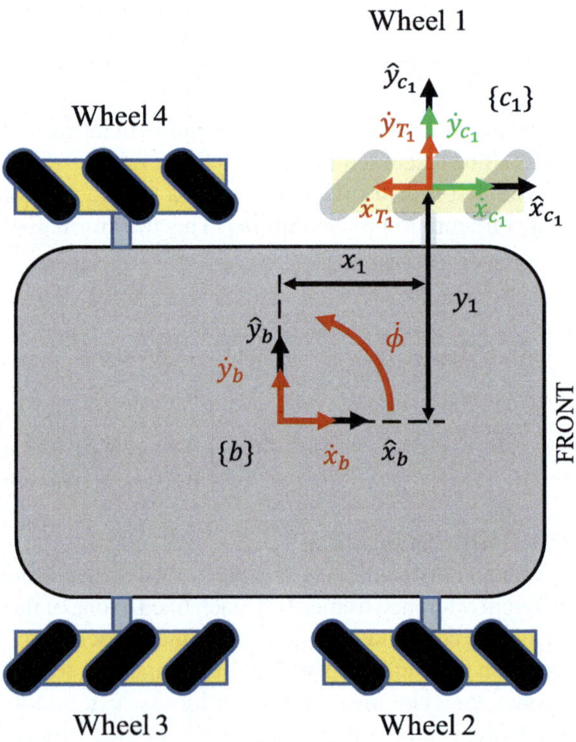

Fig. 11.3 Illustration of the transformation from the body frame $\{b\}$ to the wheel frames $\{c_i\}$. The figure shows how the chassis speeds and angular velocity contribute to calculating individual velocities in each wheel frame

wheels are distinctively equipped with rubber rollers set at an angle α relative to the rotational axis of the wheels, typically $\alpha = 45°$. These rollers, which can rotate freely upon contact with the ground, enable the wheel to effectively "slide" in the direction determined by α. This movement results from rolling, but conceptualizing it as sliding simplifies the understanding. This design grants Mecanum wheels an additional *sliding velocity*, represented as $\mathbf{v}_{\mathbf{slide}}$, which is oriented non-parallel to $\mathbf{v}_{\mathbf{motor}}$, visually illustrated in Fig. 11.4.

Referring to the diagram in Fig. 11.4, the velocities associated with Mecanum wheels, $\mathbf{v}_{\mathbf{slide}}$ and $\mathbf{v}_{\mathbf{motor}}$, are decomposed along the main axes of the reference frame $\{c_i\}$. This process of decomposition is detailed as follows [3]:

$$v_x = \|\mathbf{v}_{\mathbf{motor}}\| - \sin(\alpha) \cdot \|\mathbf{v}_{\mathbf{slide}}\| \,, \tag{11.7a}$$

$$v_y = \cos(\alpha) \cdot \|\mathbf{v}_{\mathbf{slide}}\| \,. \tag{11.7b}$$

These equations elucidate how the motor and sliding velocities of the Mecanum wheel contribute to its composite motion along the $\hat{x}$ and $\hat{y}$ directions.

An analysis of (11.7a) and (11.7b) reveals the significant advantages of Mecanum wheels. With common wheels, there is no sliding velocity, resulting in $v_y = 0$, and motion is confined to the $\hat{x}$ direction, as represented by v_x. This limitation explains why conventional vehicles cannot slide sideways, for instance, to park in a tight

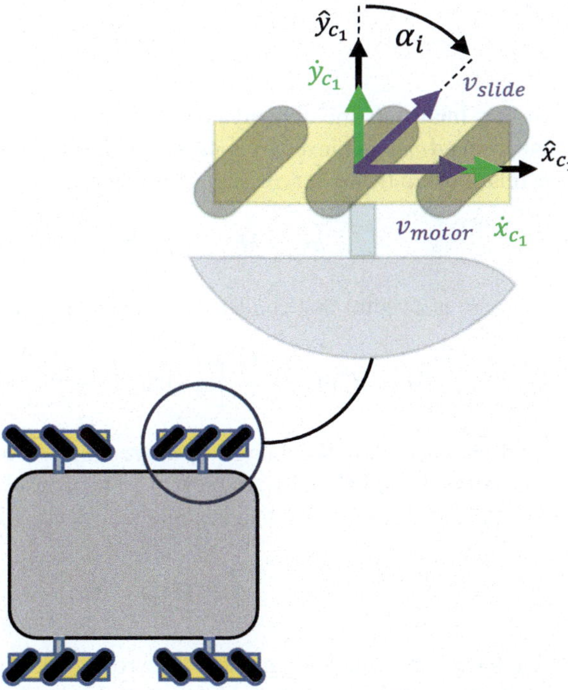

Fig. 11.4 Diagram of a Mecanum wheel showing the interplay between the orthogonal velocities $\dot{\mathbf{x}}_{c_i}$ and $\dot{\mathbf{y}}_{c_i}$ and the velocities $\mathbf{v}_{\text{motor}}$ and the sliding velocity $\mathbf{v}_{\mathbf{slide}}$

parallel spot. In contrast, Mecanum wheels allow for a portion of the frontal velocity v_x to be converted into lateral velocity v_y. By skillfully controlling these velocities, the typical movement constraints of conventional wheels are overcome, endowing the robotic platform with holonomic capabilities.

Furthermore, we can adapt the previous equations to establish a relationship that answers the question: "If I want the robot to move at a speed of v in the direction of θ (or (v_x, v_y) in rectangular coordinates), how fast must the Mecanum wheels move?" To address this, we first reconfigure the equations to express v_{motor} in terms of v_x and v_y as

$$v_{\text{motor}} = v_x + \tan(\alpha) \cdot v_y \,. \tag{11.8}$$

Considering that angular velocity ω relates to linear velocity by the equation $\omega = \frac{v}{r}$, where r is the radius of the wheel, we can substitute it into (11.8) to obtain the angular velocity of the motor as

$$\omega_{\text{motor}} = \frac{v_{\text{motor}}}{r} = \frac{v_x}{r} + \frac{\tan(\alpha) \cdot v_y}{r} \,. \tag{11.9}$$

By expressing (11.9) in matrix form and generalizing for the reference frame $\{c_i\}$, we arrive at

$$\omega_i = M_{\omega_i} \cdot \dot{\mathbf{q}}^{c_i}, \tag{11.10}$$

where the matrix M_{ω_i} is defined by

$$M_{\omega_i} = \begin{bmatrix} \frac{1}{r} & \frac{\tan(\alpha_i)}{r} \end{bmatrix}. \tag{11.11}$$

In this final step, the objective of relating the velocity configuration $\dot{q}$ with each angular speed ω_i is achieved by applying all the previously discussed transformations. The relationship that emerges is encapsulated in

$$\omega_i = M_{\omega_i} \cdot M_b^{c_i} \cdot R_s^b(\phi) \cdot \dot{q} = h_i \cdot \dot{q}. \tag{11.12}$$

Here, h_i is a vector that condenses all the matrix transformations, as defined next:

$$h_i = M_{\omega_i} \cdot M_{c_i} \cdot R_s^b(\phi) = \frac{1}{r} \cdot \begin{bmatrix} x_i \cdot \tan(\alpha_i) - y_i & \frac{\cos(\alpha_i+\phi)}{\cos(\alpha_i)} & \frac{\sin(\alpha_i+\phi)}{\cos(\alpha_i)} \end{bmatrix}. \tag{11.13}$$

Further abstraction is achieved by aggregating all motor speeds into a vector $\omega = [\omega_1\ \omega_2\ \omega_3\ \omega_4]^T$ and $\mathbf{H} = [h_1\ h_2\ h_3\ h_4]^T$. This consolidation enables the expression of (11.12) for all four wheels in a single matrix equation, as shown next:

$$\omega = \mathbf{H} \cdot \dot{q} \approx \mathbf{H} \cdot \frac{\Delta \mathbf{q}}{\Delta t}, \tag{11.14}$$

and it defines the kinematic model of the robot [7]. This equation is pivotal in determining the specific individual Mecanum wheel velocities required to achieve any desired change in the configuration of the robot, which is crucial for accurate position and speed control of the robot.

11.2.2 Odometry Model of a Mecanum Wheeled Robot

In robotics, numerous methods exist for estimating a configuration of the robot, each with its own set of advantages and limitations. One of the more straightforward methods is pose estimation through odometry. Odometry utilizes the kinematic model and measurements of wheel rotation to estimate changes in the configuration. Essentially, it provides the inverse of the relationship expressed in (11.14). By tracking the rotation of each Mecanum wheel, represented by the vector $\Delta\boldsymbol{\theta} = (\Delta\theta_1, \Delta\theta_2, \Delta\theta_3, \Delta\theta_4)$, odometry computes the resulting change in configuration $\Delta\mathbf{q}$. This enables the incremental update of the configuration as follows:

$$\mathbf{q}_{k+1} = \Delta\mathbf{q} + \mathbf{q}_k. \tag{11.15}$$

As highlighted at the beginning of Sect. 11.2.1, addressing incremental changes is more effectively done in terms of the body frame $\{b\}$. By modifying (11.12)–(11.14), a version of the kinematic model concerning the body frame $\{b\}$ can be established. This adaptation is detailed as follows:

$$\omega_i = M_{\omega_i} \cdot M_b^{c_i} \cdot \dot{\mathbf{q}}^b = h_i' \cdot \dot{\mathbf{q}}^b, \tag{11.16a}$$

$$h_i' = M_{\omega_i} \cdot M_{c_i} = \frac{1}{r} \cdot \left[x_i \cdot \tan(\alpha_i) - y_i \;\; 1 \;\; \tan(\alpha_i)\right], \tag{11.16b}$$

$$\boldsymbol{\omega} = \mathbf{H}' \cdot \dot{\mathbf{q}}^b. \tag{11.16c}$$

The objective is to solve (11.16c) for $\dot{\mathbf{q}}^b$. A straightforward approach might involve multiplying both sides by the inverse of the matrix H'. However, this is not feasible because H' is a 3×4 matrix with no inverse. The Moore–Penrose pseudoinverse, denoted as $H'^{\dagger}$, provides a viable solution. This method is advantageous because it either finds an exact solution for

$$\frac{\Delta\mathbf{q}^b}{\Delta t} \approx \dot{\mathbf{q}}^b = H'^{\dagger} \cdot \boldsymbol{\omega} \approx H'^{\dagger} \cdot \frac{\Delta\boldsymbol{\theta}}{\Delta t}, \tag{11.17}$$

if one exists, or generates a response that minimizes the error when no exact solution is possible. Considering the typical lack of movement restrictions in a Mecanum wheeled robot, finding an exact solution for (11.17) is always possible. The computation of the Moore–Penrose pseudoinverse, $H'^{\dagger}$, can be achieved through various methods, each with its own computational characteristics, as detailed in [6].

The subsequent phase involves integrating (11.17) to establish the relationship between $\Delta\boldsymbol{\theta}$ and $\Delta\mathbf{q}^b$. Screw theory provides an elegant method for this integration. According to the Chasles–Mozzi theorem [10], any rigid-body motion, such as $\dot{\mathbf{q}}^b$ occurring in a time interval Δt, can be represented as a rotation about an axis combined with a translation along the same axis. In essence, it proposes that any displacement of a rigid body can be conceptualized as a screw motion. This motion is characterized by two key elements: a screw axis and a rotational component. The screw axis is denoted by the vector $S_{6\times 1} = [\hat{\boldsymbol{\psi}} \;\; \mathbf{v}_u]^T$, consisting of the unit rotation axis ($\hat{\boldsymbol{\psi}} = (\psi_x, \; \psi_y, \; \psi_z)$) and a corresponding "unitary" velocity vector ($\mathbf{v}_u = (v_{u_x}, \; v_{u_y}, \; v_{u_z})$). Conversely, the rotational component is specified by a scalar element, $\Delta\psi$.

Consequently, to express $\dot{\mathbf{q}}^b$ using screw theory notation, it is necessary to expand the vector $\dot{\mathbf{q}}^b = (\dot{\phi}, \; \dot{x}, \; \dot{y})$ from a three-parameter vector to a sixth-dimensional representation as shown below

$$\dot{\mathbf{q}}^b = \begin{bmatrix} \dot{\boldsymbol{\phi}} \\ \dot{\mathbf{r}} \end{bmatrix} = \begin{bmatrix} 0 \\ 0 \\ \dot{\phi} \\ \dot{x} \\ \dot{y} \\ 0 \end{bmatrix} = S \cdot \Delta\psi = \begin{bmatrix} \hat{\boldsymbol{\psi}} \\ \mathbf{v}_u \end{bmatrix} \cdot \Delta\psi. \tag{11.18}$$

To ensure the unitary property of the unitary screw axis $\hat{\boldsymbol{\psi}}$, it is necessary to normalize $\dot{\boldsymbol{\phi}}$ as shown next:

$$\hat{\boldsymbol{\psi}} = \frac{\dot{\boldsymbol{\phi}}}{\Delta\psi}, \tag{11.19}$$

where $\Delta\psi$ is defined as

$$\Delta\psi = \begin{cases} \|\dot{\mathbf{r}}\| & \text{if } \|\dot{\boldsymbol{\phi}}\| = 0, \\ \|\dot{\boldsymbol{\phi}}\| & \text{if } \|\dot{\boldsymbol{\phi}}\| \neq 0, \end{cases} \tag{11.20}$$

to handle the case when $\|\dot{\boldsymbol{\phi}}\| = 0$ and avoid division by 0. Note that Eq. (11.19) also defines the "unitary" velocity vector as

$$\mathbf{v}_u = \frac{\dot{\mathbf{r}}}{\Delta\psi}. \tag{11.21}$$

Now, it is a good time to recall that both $\dot{\mathbf{r}}$ and $\dot{\boldsymbol{\phi}}$ are obtained through the operation $H'^{\dagger} \cdot \Delta\boldsymbol{\theta}/\Delta t$ and depend solely on the vector $\Delta\boldsymbol{\theta}$, which stores the rotational change of the wheels. Consequently, $\Delta\psi$, $\hat{\boldsymbol{\psi}}$, and $\mathbf{v}_u$ also depend exclusively on $\Delta\boldsymbol{\theta}$.

Utilizing screw theory rotation, the matrix exponential, denoted as $e^{[S]\Delta\phi}$, is applied as an integrator [9]. This operation is described as

$$T_{b'}^{b} = e^{[S]\Delta\psi} = \begin{bmatrix} R_{b'}^{b} & p_{b'}^{b} \\ \mathbf{0} & 1 \end{bmatrix} = \Delta\mathbf{q}^{b}, \tag{11.22}$$

where $R_{b'}^{b}$ represents a 3×3 rotation matrix known as the Rodrigues formula [7], and $p_{b'}^{b}$ is a 3×1 vector. Both $R_{b'}^{b}$ and $p_{b'}^{b}$ are defined by piecewise equations

$$R_{b'}^{b} = \begin{cases} I & \text{if } \|\dot{\psi}\| = 0, \\ I + \sin(\Delta\psi) \cdot [\hat{\psi}] + (1 - \cos(\Delta\psi)) \cdot [\hat{\psi}]^2 & \text{if } \|\dot{\psi}\| \neq 0, \end{cases} \tag{11.23a}$$

$$p_{b'}^{b} = \begin{cases} \mathbf{v}_u \cdot \Delta\psi & \text{if } \|\dot{\psi}\| = 0, \\ (I \cdot \Delta\psi + (1 - \cos(\Delta\psi)) \cdot [\hat{\psi}] \\ \quad + (\Delta\psi - \sin(\Delta\psi)) \cdot [\hat{\psi}]^2) \cdot \mathbf{v}_u & \text{if } \|\dot{\psi}\| \neq 0 \end{cases} \tag{11.23b}$$

to handle the case where $\|\dot{\phi}\| = 0$, analogous to Eq. (11.20). Notice that this scenario occurs when the robot undergoes purely linear motion without rotational movement. Conversely, the case $\|\dot{\phi}\| \neq 0$ accounts for scenarios involving rotation and/or displacement.

It is important to note that $R_{b'}^{b}$, $p_{b'}^{b}$, and consequently, $\Delta\mathbf{q}^{b} = T_{b'}^{b}$ are solely dependent on $\Delta\boldsymbol{\theta} = (\Delta\theta_1, \Delta\theta_2, \Delta\theta_3, \Delta\theta_4)$. Equation (11.22) thus forms the final component necessary to estimate the configuration through incremental updates, applying (11.15). These updates are based solely on the rotation of the wheels $\Delta\boldsymbol{\theta}$, measured by the motor encoders over a time interval Δt.

11.2.3 Visibility Graphs for Mapping the Free Space of the Environment

As mentioned before, the appropriate description of the environment is crucial. This description must encompass a definition of free space, the area where the robot can move without colliding or stepping out of the boundaries of the environment. Additionally, this description must be in a format that enables algorithms to explore the free space and devise paths effectively. Therefore, in warehouse-like environments, where a layout detailing the locations of shelves and products is available, employing a visibility graph is an effective strategy for defining the free space.

Visibility graphs define the environment by placing nodes at strategic locations, such as the corners of zones that the robot should avoid, known as the non-free space, as well as at other points of interest [5]. Edges are then formed between pairs of nodes with a “clear line of sight”, thus the term visibility graph. In Fig. 11.5, a simple environment layout and its corresponding visibility graph are illustrated. Nodes are positioned at the corners of off-limit zones, primarily around obstacles like shelves or areas beyond the boundaries of the environment, constituting the non-free

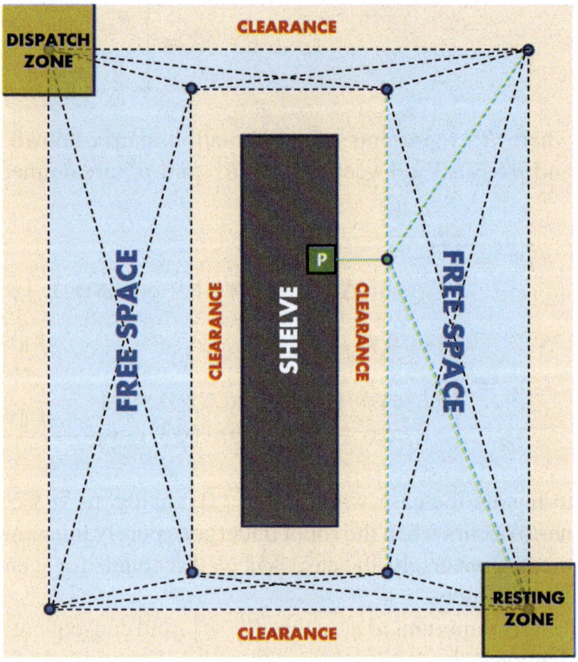

Fig. 11.5 A simplified depiction of a warehouse layout featuring a single shelf and two aisles, accompanied by its corresponding visibility graph. The nodes are represented by colored circles, with edges illustrated as dashed lines. The free-space areas are highlighted in blue, and a point of interest, product "P", is indicated in green, showcasing its inclusion in the visibility graph

space. Note that, in Fig. 11.5, nodes are strategically placed not exactly at the corners of the non-free space; they are set to provide necessary clearance, considering the width of the robot. An edge is established whenever a straight line between two nodes does not intersect any non-free space. When applied across all node pairs, this method results in a graph-based representation of the free space. Once the free space is defined using the visibility graph, any point of interest can be seamlessly integrated into the graph, allowing for efficient path computation to these locations. In Fig. 11.5, a single point of interest, labeled as product "P" and shown in green, is incorporated. Green dashed lines represent the additional edges in the visibility graph to integrate this new node of interest. These edges allow the path-finding algorithm to explore the graph and formulate an optimal trajectory to retrieve the specified product.

11.2.4 Prototype and Digital Twin System Description

This subsection integrates the previously discussed models to provide a comprehensive overview of the systems for both the robot prototype and the digital twin benchmarks. Figures 11.6 and 11.7 show that the benchmark system architecture is segmented into various blocks, each delineating a distinct component.

Beginning with the prototype benchmark depicted in Fig. 11.6, the general inputs for the system include a map of the warehouse layout and a list of products designated

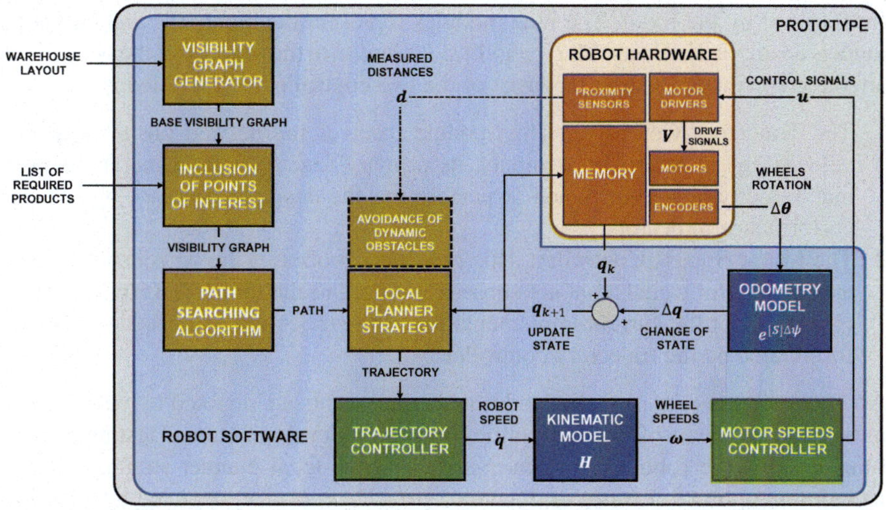

Fig. 11.6 System architecture diagram of the robot prototype benchmark. The diagram showcases the integration of various modules, including the kinematic and odometry models (blue), trajectory and speed controllers (green), and additional components related to the cognitive capabilities (yellow)

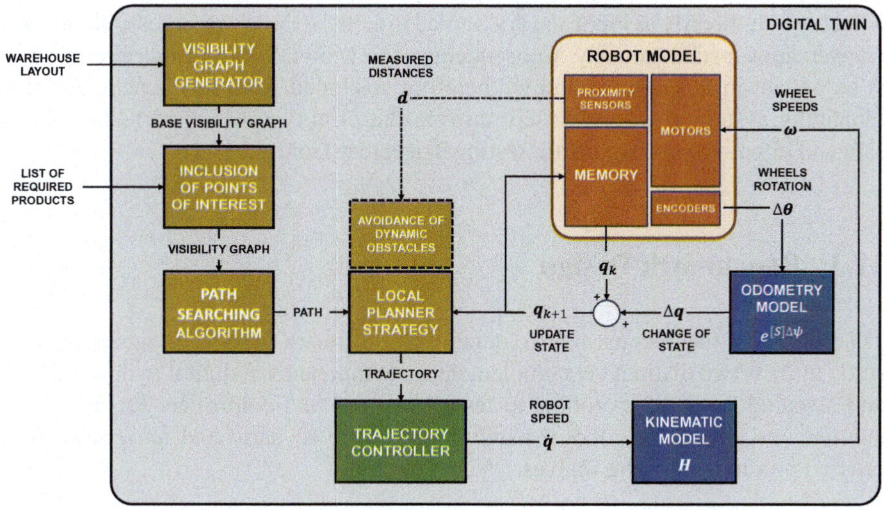

Fig. 11.7 System architecture of the digital twin benchmark. This schematic representation highlights the integration of the various modules within the digital twin system. The blue modules illustrate the kinematic and odometry models, the green modules focus on the trajectory controllers, and the yellow modules indicate additional cognitive components of the robot system

for retrieval by the robot. The blue modules represent the kinematic and odometry models developed in Sects. 11.2.1 and 11.2.2. Central to the purpose of the benchmark are the green modules, which consist of the two controllers to be tested:

1. The Trajectory Controller: This module receives the desired position ($\mathbf{q}$) and velocity ($\dot{\mathbf{q}}$) of the robot as inputs. Its primary function is to generate an optimal velocity vector $\dot{\mathbf{q}}$, aimed at performing the desired trajectory by selected performance indicators.
2. The Motor Speeds Controller: This controller processes the wheel speeds, generating control signals that are subsequently fed to the motor drivers. This stage is integral for the current implementation of the trajectory and speed commands formulated by the trajectory controller.

Apart from these core modules, additional components are denoted by yellow modules. These elements offer opportunities for further development and testing of algorithms within the robotics domain. Notice that the local planner strategy module includes an optional feature for dynamic obstacle avoidance, depicted by a module with dashed lines.

Continuing with the discussion, Fig. 11.7 presents the system architecture for the digital twin benchmark. As expected, it resembles the prototype benchmark, with the notable difference being the substitution of the robot hardware with an idealized robot model in the digital twin. This model is ideal as it is programmed to receive the desired wheel speeds directly as input and is assumed to achieve these velocities without error or overshoot instantaneously. Consequently, the Motor Speeds Controller module is redundant in this context and is, therefore, excluded from this benchmark. The remaining system structure closely mirrors the prototype benchmark, facilitating safe and effective simulations for testing Trajectory Controllers.

11.3 Benchmark Design

This section of the chapter aims to detail the benchmark design. As mentioned, the study utilizes two distinct yet complementary components: a digital twin simulation and a scaled low-cost prototype to test algorithms and controllers for driving an autonomous holonomic robot through a warehouse scenario and retrieve a list of ordered products from the shelves.

11.3.1 Digital Twin Benchmark

The experimental setup encompasses several core modules, as detailed in Fig. 11.7, that collectively simulate the functioning of the robot within a virtual environment. Central to this setup are the kinematic and odometry models. These models are vital for accurately replicating the movements, ensuring that the simulation mimics the real-world dynamics of the motion of the robot.

Complementing these fundamental models are what we refer to as "cognitive modules". These include an algorithm for generating visibility graphs based on the input warehouse layout and a list of products for retrieval. This algorithm plays a crucial role in mapping the environment for path planning. Additionally, the setup integrates a path-finding algorithm, particularly the A* algorithm combined with Brute Force search. This combination is instrumental in optimizing routes during scenarios that involve retrieving multiple products, closely simulating complex navigation challenges encountered in real-world applications.

The digital twin also simulates the distance sensors, which provide essential data regarding the proximity of obstacles, thereby aiding the ability to detect potential obstacles in its trajectory. This navigation is facilitated through the local planner module, grounded in Virtual Potential Fields. It allows the robot to adapt its path in response to environmental obstacles, even when they were not defined in the warehouse layout. For further insights into the robotics algorithms implemented, as well as additional concepts, readers can refer to the comprehensive resource [11].

A simplified model of the robot is a pivotal element of the setup of the digital twin. This model is implemented to process the desired wheel speeds as input and, based on a user-adjustable time parameter (Δt), output the rotations of the wheels, akin to what would be expected from a real wheel encoders. This simplified model operates under idealized conditions. Such a level of abstraction renders the motor speeds controller module unnecessary for this benchmark, simplifying the simulation framework.

The test environment within the digital twin benchmark is a simulated warehouse based on the predefined layout. This virtual setting allows users to experiment with diverse warehouse sizes and designs tailored to test the effectiveness of their control systems under varying conditions. A notable feature of this environment is the ability to integrate unforeseen obstacles into the predefined layout. Although not part of the original layout used to generate the visibility graph, these obstacles are detectable by the simulated distance sensors of the digital twin. This capability is crucial for evaluating trajectory controllers and local planners, particularly when the robot encounters unknown obstacles. Users are thus encouraged to push the boundaries of their controllers and algorithms, enhancing the ability of the robot to navigate and perform in unanticipated situations. The simulated warehouse environment serves as an ideal and secure platform, providing a controlled setting favorable for the initial testing and development stages.

The digital twin benchmark, despite its utility, inherently possesses limitations due to the simplification of real-world complexities into a more manageable virtual environment. A critical constraint is the inability of the digital twin to replicate the intricate real-world challenges faced by a robot, such as dealing with acceleration and velocity constraints, sensor noise from devices like ultrasonic sensors and rotary encoders, and communication delays and latency among interacting electronic modules. These real-world intricacies, abstracted in the simulation, restrict the range of scenarios where controllers can be rigorously tested. However, this abstraction also has an upside. It provides a streamlined and straightforward platform for testing new controllers and algorithms. This simplification allows for rapid and safe experimentation, speeding up the design, implementation, testing, and analysis cycle. The digital twin thus serves as an effective initial step in the development process, enabling quick identification of potential improvements and optimizations before advancing to more complex real-world applications.

11.3.2 Scaled Prototype Benchmark

The prototype benchmark is designed for modularity, cost-efficiency, and space-saving attributes. Its mechanical structure is particularly designed for additive manufacturing, commonly known as 3D printing. It features four mecanum wheels, each powered by an independent DC motor equipped with a magnetic rotary encoder to measure wheel rotations. The prototype includes an array of sixteen ultrasonic sensors encircling the robot, enabling it to accurately measure distances to nearby objects. A simple gantry mechanism is installed atop the robot for product retrieval in a scaled warehouse setting.

The electronic hardware of the prototype centers around a Raspberry Pi Zero, which functions as the main control board. Additionally, it integrates four Atmega 328-P microcontrollers, each dedicated to controlling the speed of one motor.

Software-wise, the prototype benchmark encompasses all necessary system modules, illustrated in Fig. 11.6, mirroring the implementations found in the digital twin. However, in this benchmark, the Motor Speeds Controller module is a critical addition that is pivotal for precise velocity control. Nonetheless, the setup offers the flexibility to employ a pre-implemented PID controller for the Motor Speeds Controller, which is particularly beneficial when the primary focus is testing Trajectory Controllers without delving into motor speed intricacies.

The testing environment for the prototype benchmark must replicate a scaled-down warehouse. A key requirement for this setup is a flat, non-slippery surface, aligning with the design specifications unsuitable for uneven terrain. For safety and practicality, it is advisable to use cardboard constructions to simulate shelves, minimizing potential damage to both the robot and the environment during collisions, as shown in Fig. 11.8. Additionally, a safety button is incorporated into the prototype design, allowing for immediate cessation of operations when necessary.

The prototype benchmark poses a more complex and realistic challenge. It necessitates the implementation of algorithms and controllers capable of operating in a noisy environment, where sensors and actuators behave less than ideally. One significant challenge is designing efficient controllers and algorithms that minimize latency, thereby ensuring prompt and accurate responses from the robot.

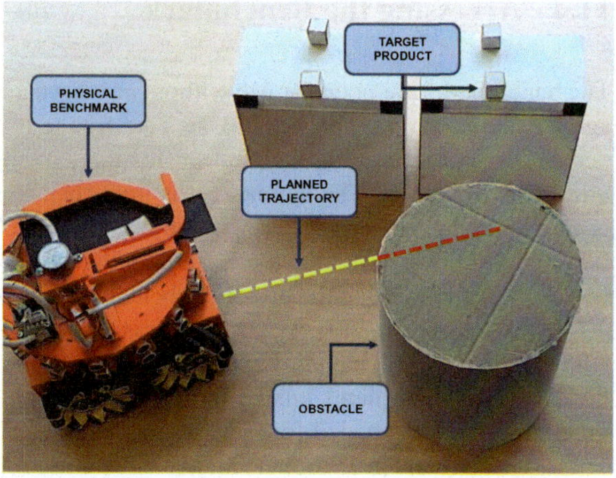

Fig. 11.8 Image of the physical benchmark setup. It includes the scaled prototype moving toward the target product for retrieval and an obstacle impeding the planned trajectory

11.3.3 Data Collection and Performance Metrics

The core purpose of benchmarks is to facilitate the comparison of new controller or algorithm solutions with previously tested implementations. This requires thorough data collection and clearly defined performance metrics. This section outlines essential data and suggests performance indicators for both the digital twin and prototype benchmarks.

First, data gathered in both benchmarks is the configuration of the robot ($\mathbf{q} = (\phi, x, y)$), accompanied by timestamps. This data is vital for comprehending the movement within the environment. Additionally, unique to the prototype benchmark, encoder data is collected, which records each rotational speed and displacement of the wheel. This data is also timestamped to document the wheel speeds at each moment meticulously.

Furthermore, we present some performance metrics to evaluate controllers and algorithms in the benchmarks:

- Elapsed time for product retrieval: measured in seconds.
- Path search time: recorded in seconds.
- Path accuracy and deviation: measured in distance units (centimeters for the scaled prototype) or percentage.
- Obstacle avoidance efficiency: number of successful obstacle avoidances.
- System responsiveness: measured in milliseconds.
- Robustness: testing the controller in different warehouse layouts.

These metrics provide a framework for analyzing the strengths and limitations of controllers and algorithms being tested.

11.4 Accessing the Benchmark

The benchmarks are open for use without any licensing restrictions or costs, encouraging adaptation and innovation. Users are free to modify the code and mechanical designs or to use them as inspiration for their own projects. For those aiming to utilize the digital twin benchmark, it is important to note that MATLAB® R2022b is a prerequisite. Access to this specific version of MATLAB® is essential to ensure full compatibility and the smooth functioning of the benchmark. Additionally, the CAD designs integral to the prototype benchmark were developed using SolidWorks 2022. Therefore, access to this software is necessary for users who plan to engage with the prototype benchmark, particularly in modifying or working with the CAD files. These requirements ensure that users can seamlessly interact with and adapt the benchmarks as needed for their purposes.

Researchers and practitioners interested in utilizing the benchmarks for their work can access all software resources related to both the digital twin and prototype benchmarks via the GitHub repository. The repository can be found at https://github.com/mateovernaza/mecanum-wheeled-robot-benchmarks.git. This repository hosts the complete suite of software tools, algorithms, and simulations required for effectively using the benchmarks. For those looking to build and test the physical prototype, the necessary CAD files and instructions are available on GrabCAD. The complete set of CAD-designed files can be downloaded from https://grabcad.com/library/mecanum-wheeled-robot-prototype-benchmark-1. The hands-on experience obtained by manufacturing and assembling the prototype will offer invaluable insights into the functionality and inner workings. The process, while demanding, provides a deeper understanding of the complexities involved in robotics. For a more comprehensive understanding of the benchmark systems and their applications, interested users can refer to [13].

11.5 Discussion

This chapter presents an innovative approach to robotics benchmarking, focusing on developing a digital twin and a scaled low-cost prototype for autonomous holonomic robots. This approach represents a valuable contribution to the field by addressing the requirement for effective and efficient testing of control algorithms. The digital twin replicates the environment and behavior of the robot, enabling the testing of control systems under different simulated conditions. Meanwhile, the scaled prototype is crucial for assessing real-world challenges such as communication, environmental disruptions, and design uncertainties, particularly in warehouse-type environments.

This approach has a key strength in its ability to quickly and cost-effectively iterate control systems in a simulated environment, which is essential in the early stages of development. The digital twin, in particular, provides a controlled setting for initial testing and analysis, which accelerates the design and implementation cycle.

However, these methods also have limitations. For example, the digital twin may not fully capture all real-world complexities, especially those involving sensor noise and physical system constraints. Additionally, while more realistic, scaled prototypes may still struggle to replicate full-scale operational environments comprehensively.

In robotics, these methods contribute to research using simulation and scaled models for system development and testing. For example, works like [2] and [12] demonstrate similar concepts in different application areas, showing the increasing trend in using digital twins and prototypes to enhance accuracy and efficiency.

Despite advancements, there are gaps in simulating real-world conditions and scaling prototypes. To address these, researchers are exploring advanced simulation software, AI-based predictive models, sophisticated sensor technologies, and machine learning algorithms in robotics research.

References

1. Ahn M, Zhu H, Hartikainen K, Ponte H, Gupta A, Levine S, Kumar V (2020) Robel: robotics benchmarks for learning with low-cost robots. In: Proceedings of the conference on robot learning. Virtual, . PMLR, pp 1300–1313
2. Aranda Joaquín (2006) Pablo González de Santos, and Jesús de la Cruz. Robotics and Automation in the Maritime Industries, Producción Gráfica Multimedia
3. Brown N, Pierce T, Ma G, McCusker J, Cuckov F (2022) Development of kinematic and dynamic model of an omnidirectional four Mecanum wheeled robot. In: Proceedings of the ASME international mechanical engineering congress and exposition, vol 86670. Columbus, Ohio. American Society of Mechanical Engineers, p V005T07A031
4. Calisi D, Iocchi L, Nardi D (2008) A unified benchmark framework for autonomous mobile robots and vehicles motion algorithms (MoVeMA benchmarks). In: Proceeding of the workshop on experimental methodology and benchmarking in robotics research (RSS 2008). Zurich, Switzerland
5. Contreras JD, Martínez F, Martínez FH (2015) Path planning for mobile robots based on visibility graphs and A* algorithm. In: Proceedings of the seventh international conference on digital image processing (ICDIP 2015), vol 9631. Los Angeles, USA. International Society for Optics and Photonics, SPIE, p 96311J
6. Klimczak Marek, Cecot Witold, González David (2019) On Moore-Penrose pseudoinverse computation for stiffness matrices resulting from higher order approximation. Math Probl Eng 2019:5060397
7. Lynch KM, Park FC (2017) Modern robotics. Cambridge University Press
8. Muñoz Nelson, Valencia Jaime, Londoño Nelson (2010) Quantitative performance metrics for mobile robots navigation. In: Barrera Alejandra (ed) Mobile robots navigation, chapter 24. IntechOpen, Rijeka
9. Savcı ÜZ (2022) Screw motion via matrix algebra in three-dimensional generalized space. Symmetry 14(11)
10. Selig JM (2004) Lie groups and lie algebras in robotics. In: Byrnes J (ed) Proceedings of the computational noncommutative algebra and applications. Dordrecht. Springer Netherlands, pp 101–125
11. Siciliano B (2016) Springer handbook of robotics. Springer Cham
12. Söderberg R, Wärmefjord K, Carlson JS, Lindkvist L (2017) Toward a digital twin for real-time geometry assurance in individualized production. CIRP Annals 66(1):137–140
13. Vernaza Ortiz M (2022) Diseño de un robot asistencial para la compra en un supermercado. Universidad Internacional del Ecuador

Chapter 12
Benchmarking Handling Performance in 4WD Electric Vehicles Through Advanced Control Techniques

Ahmed Hassan, Sara Ruiz-Moreno, Jose Ramon D. Frejo, and José M. Maestre

12.1 Introduction

This chapter introduces a benchmark for a real car at the University of Seville as illustrated in Fig. 12.1, focusing on overcoming challenges in Electric Vehicle (EV) technology development. The benchmark's main goal is to assess specific aspects of EV performance, particularly lateral stability, and maneuverability during the double-lane change maneuver. It emphasizes the use of various advanced control techniques to address issues like oversteering and understeering during cornering maneuvers, while also addressing real-time constraints imposed by advanced software functionalities. The EV platform provides opportunities for refining motion control strategies, employing precise motor torque control and emerging technologies like In-Wheel Motors (IWMs) for optimized space utilization and independent wheel control [12]. In particular, the four-Wheel Independent-Motor-Drive Electric Vehicle (4WIMD EV) configuration present in Fig. 12.1 offers significant opportunities for implementing diverse control algorithms, enabling independent torque adjustments for each wheel and enhancing the vehicle's dynamic capabilities. This setup has emerged as an attractive research area in recent decades due to its environmental benefits and its capacity to adjust vehicle motion through the independent distribution of driving and braking torques on each wheel [5]. Proposed controllers

A. Hassan (✉) · S. Ruiz-Moreno · J. R. D. Frejo · J. M. Maestre
University of Seville, 41092 Seville, Spain
e-mail: ahmhasahm@alum.us.es

S. Ruiz-Moreno
e-mail: srmoreno@us.es

J. R. D. Frejo
e-mail: jdominguez3@us.es

J. M. Maestre
e-mail: pepemaestre@us.es

J. M. Maestre and C. Ocampo-Martinez (eds.), *Control Systems Benchmarks*, Advances in Industrial Control, https://doi.org/10.1007/978-3-031-76312-0_12

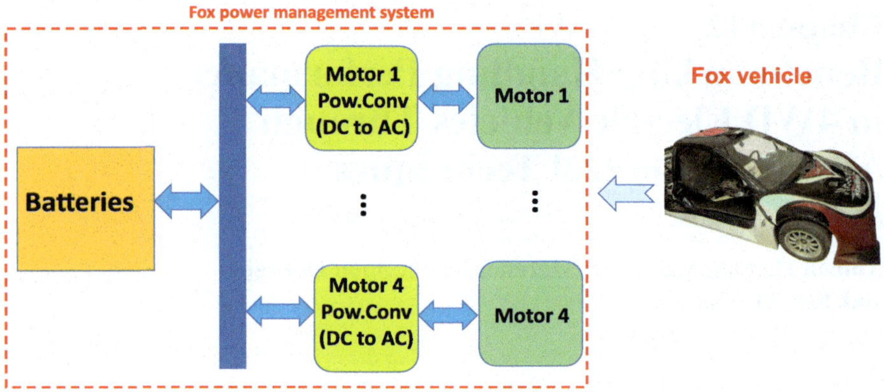

Fig. 12.1 The Fox 4WIMD EV

can be evaluated using a simulated vehicle equipped with four electric motors on its wheels, based on a real-life prototype, serving as a test platform for control strategies aimed at enhancing vehicle stability and motion control, thereby contributing to a more sustainable transportation sector.

Our benchmark is aligned with the shift toward EVs in automotive manufacturing, driven by their environmental advantages and decreased dependency on traditional fuel sources, which has underscored the necessity for precise assessment tools within control engineering. By expanding benchmark features and integrating advanced technologies such as Electronic Stability Control (ESC), Anti-lock Braking Systems (ABS), Collision Avoidance Systems, Adaptive Cruise Control (ACC), Lane Departure Warning (LDW), and Emergency Brake Assist (EBA), alongside passive safety systems, the automotive industry can also reduce injury severity and establish a safer vehicle environment; therefore this comprehensive approach not only addresses safety concerns but also contributes to improving energy efficiency and enhancing overall vehicle stability, aligning with the development of automotive engineering and consumer expectations. For these reasons, the benchmark serves as a valuable tool for evaluating these control strategies, enabling researchers to refine and validate their approaches. It impacts not only the development of EVs but also autonomous vehicles, where precise control of vehicle dynamics is crucial for safe and efficient operation, leading to safer and more efficient transportation systems.

12.2 Problem Description

The benchmark is designed to address the challenge of evaluating lateral stability enhancement during a double-lane change maneuver in vehicles, particularly focusing on addressing issues related to oversteering and understeering. The system under control is a 4WIMD EV which is utilized for the application of advanced control techniques. The primary goal of the control system is to enhance vehicle stability and maneuverability during the double-lane change maneuver, ensuring safe and smooth navigation through corners. During the double-lane change maneuver, the vehicle is subjected to rapid lateral acceleration, which can lead to instability and loss of control if not properly managed. Oversteering occurs when the vehicle turns more sharply than intended, causing the rear end to slide outwards, while understeering happens when the vehicle fails to turn as much as intended, resulting in a wider turning radius than desired, as seen in Fig. 12.2. Both of these phenomena pose significant safety risks and can lead to accidents if not corrected efficiently. The control system aims to address these challenges by utilizing advanced techniques.

Traditional stability control systems, like ESC-based brakes, often exhibit limitations in their ability to efficiently manage vehicle dynamics, as they primarily target specific tires, leading to aggressive actions and significant speed changes. To overcome these limitations, we integrate ESC with Torque Vector Control (TVC) as depicted in Fig. 12.2. Unlike ESC-based brakes, ESC-based TV utilizes TVC to engage all vehicle wheels for corrective action, thereby improving efficiency. By adjusting torque symmetrically, it maintains the driver's desired speed, minimizing abrupt speed changes. Additionally, the combination of Direct Yaw Control (DYC) with TVC further enhances vehicle stability and maneuverability: DYC creates rotational motion around the vehicle's vertical axis, while TVC optimizes torque distribution to individual wheels, effectively addressing oversteering and understeering to ensure safe and precise vehicle operation, even in challenging driving conditions. By implementing DYC and TVC, the control system also manages oversteering and understeering during the double-lane change maneuver effectively, ensuring that the vehicle maintains its intended trajectory without loss of control. The benchmark provides a standardized methodology for evaluating the performance of control techniques in this context (Fig. 12.3).

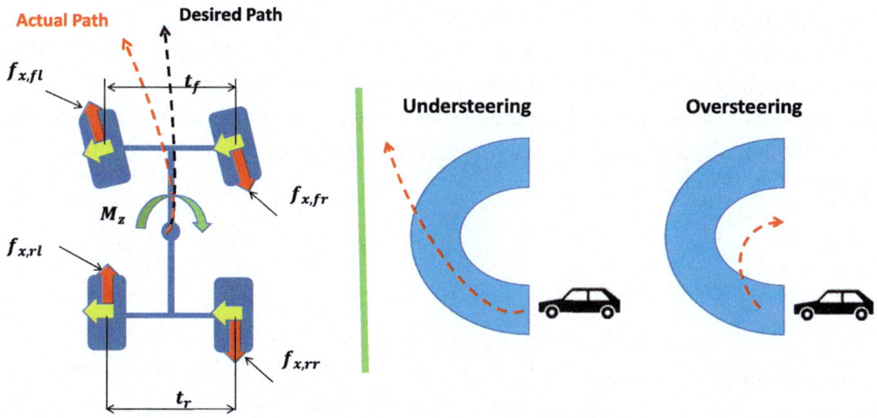

Fig. 12.2 TVC, oversteering and understeering

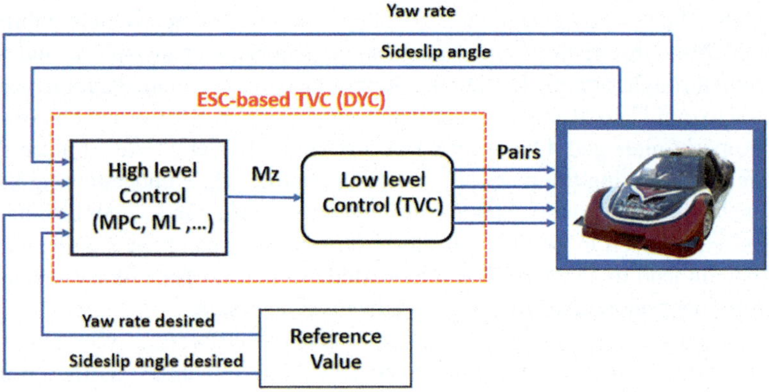

Fig. 12.3 Control structure

12.2.1 System Modeling

The mathematical model, developed within ADAMS-Car, a comprehensive virtual prototype software for modeling, equation solving, optimization, simulation, and visualization, simulates the dynamic behavior of the vehicle. SimMechanics, a Simulink® toolbox specializing in modeling mechanical systems through body and joint blocks, is utilized [11]. The Fox vehicle's model (originally imported from SolidWorks CAD) comprises 35 bodies and 38 joints, and considers 14 degrees of freedom, with simplifications such as considering rigid bodies and factoring in air friction along the longitudinal axis as is shown in Fig. 12.4. SimMechanics offers a versatile framework for in-depth analysis of vehicle dynamics. Furthermore, it accurately replicates the vehicle's trajectory and serves as both an internal model for controllers and as a digital twin of the physical vehicle with a 10 ms sampling time. Initially, a two-degree-of-freedom vehicle model focusing on yaw and lateral motion is introduced. See Table 12.1 for a list of the main variables and parameters included in the modeling process. The following state-space model is derived employing the kinematic bicycle model for lateral dynamics to facilitate control analysis assuming that a moment M_z can be produced through the difference in longitudinal tire forces between the wheels [4], i.e.,

$$\dot{x}_l = Ax_l + B_\delta \delta_f + B_M M_z, \tag{12.1}$$

where

$$x_l = \begin{bmatrix} \beta \\ \dot{\beta} \\ r \\ \dot{r} \end{bmatrix}, \quad A = \begin{bmatrix} 0 & 1 & 0 & 0 \\ -\frac{C_f+C_r}{\tau_{lag} m v_x} & -\frac{1}{\tau_{lag}} & \left(\frac{C_r l_r - C_f l_f}{\tau_{lag} m v_x^2} - \frac{1}{\tau_{lag}}\right) & -1 \\ 0 & 0 & 0 & 1 \\ \frac{C_r l_r - C_f l_f}{\tau_{lag} I_z} & 0 & -\frac{C_f l_f^2 + C_r l_r^2}{\tau_{lag} I_z v_x} & -\frac{1}{\tau_{lag}} \end{bmatrix}, \tag{12.2}$$

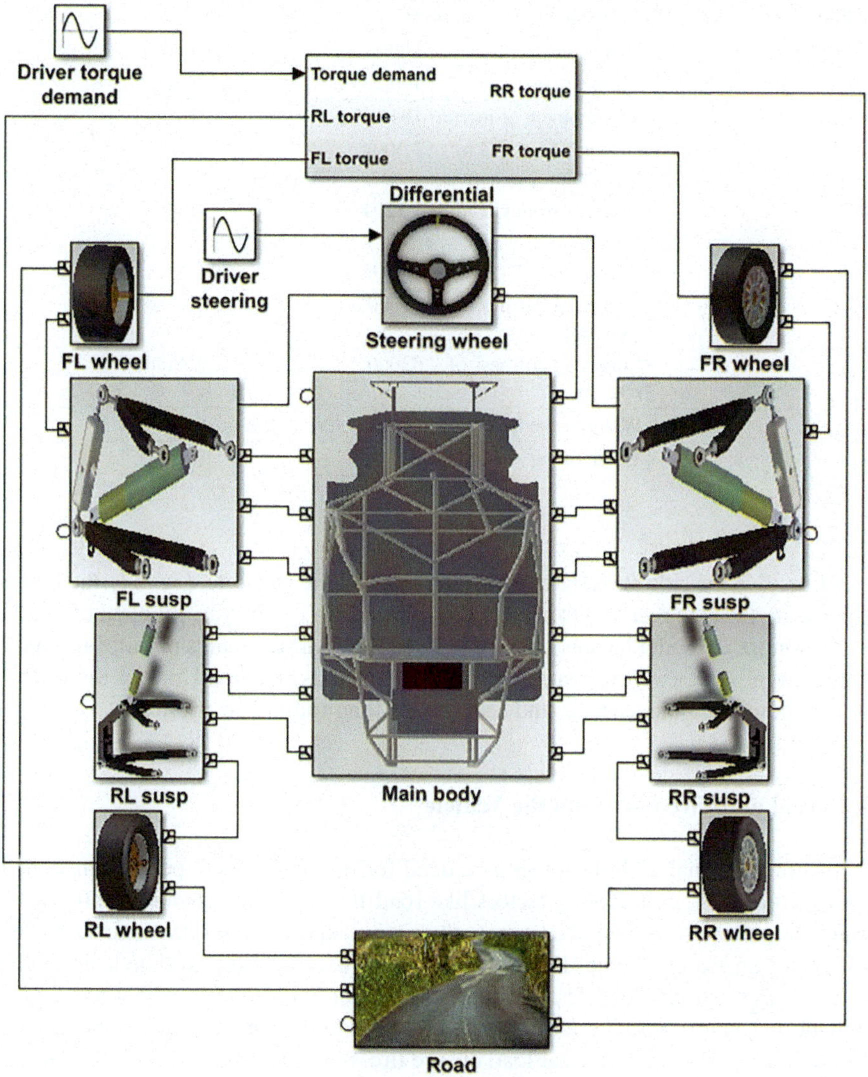

Fig. 12.4 The vehicle's SimMechanics model

$$B_\delta = \begin{bmatrix} 0 \\ \frac{C_f}{\tau_{lag} m v_x} \\ 0 \\ \frac{C_f l_f}{\tau_{lag} I_z} \end{bmatrix}, B_M = \begin{bmatrix} 0 \\ 0 \\ 0 \\ \frac{1}{\tau_{lag} I_z} \end{bmatrix}. \tag{12.3}$$

Table 12.1 Vehicle parameters and model variables

Symbol	Description	Value	[Units]
m	Vehicle mass	400.238	[kg]
I_z	Yaw moment of inertia	1047.52	[kg m^2]
l_f	Distance from CG to front axle	1.3004	[m]
l_r	Distance from CG to rear axle	1.2204	[m]
L	Wheelbase	2.5208	[m]
C_f	Cornering stiffness of front tire	47000	[N/rad]
C_r	Cornering stiffness of rear tire	53000	[N/rad]
t_f	Width of the front axle	1.55	[m]
t_r	Width of the rear axle	1.48	[m]

Here, β is the side slip angle, m is the mass, r is the yaw rate, l_r and l_f the Center of Gravity (CG) to rear and front axle distances, and I_z is the yaw moment of inertia.

A nonlinear model has also been developed in SimMechanics to faithfully replicate the real vehicle. We also assume rigid vehicle parts, tire–road force at the wheel's lowest point, and flat surface, and account for longitudinal air friction.

External and Tire Forces on the Vehicle

Managing external and tire forces is crucial for optimal vehicle performance, handling, and safety, considering factors like road friction, aerodynamics, and gravity. Balancing these forces ensures a smooth driving experience while maximizing fuel efficiency and safety. Figure 12.5a illustrates external forces on the vehicle, including gravity (F_g), vertical load (F_z), rolling resistance (F_roll), road-to-wheel force (F_w), and air resistance (F_air). Gravity acts proportional to the vehicle's mass, and gravitational acceleration and vertical load ensure tire–road alignment without violations, i.e.,

$$F_\mathrm{g} = -mg. \tag{12.4}$$

Here, F_roll denotes non-elastic tire force, partly dissipated upon pressure release, i.e.,

$$F_\mathrm{roll} = f_\mathrm{roll} F_\mathrm{z}. \tag{12.5}$$

The displacement of air during forward motion gives F_air, dependent on frontal area (S_front), air density (ρ_air), and longitudinal speed (v_x), with a shape-dependent drag coefficient (C_ax), i.e.,

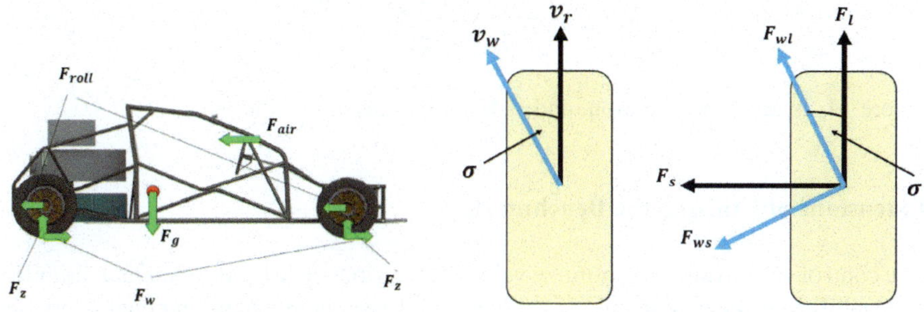

(a) External forces. (b) Tire's forces.

Fig. 12.5 External and tire forces on the vehicle

$$F_{\text{air}} = 0.5 C_{\text{ax}} S_{\text{front}} P_{\text{air}} v_{\text{x}}^2. \tag{12.6}$$

Figure 12.5b depicts tire–road interaction, governed by longitudinal (S_L) and lateral (S_s) coefficients, influencing forces at the contact surface, which are described as follows:

$$S_{\text{L}} = \frac{v_{\text{r}} \sin \alpha}{v_{\text{w}}}, \quad S_{\text{S}} = \frac{v_{\text{r}} \cos \alpha - v_{\text{w}}}{v_{\text{r}} \cos \alpha}, \tag{12.7}$$

where v_{w} is the wheel speed, v_{r} is a virtual speed aligned with the x_i axis, and α is for the slip angle between v_{r} and v_{w}. The longitudinal and lateral forces (F_{L}, F_{S}) are

$$F_{\text{L}} = \mu_{\text{L}} F_{Z}, \, F_{\text{S}} = \mu_{\text{S}} F_{Z}, \tag{12.8}$$

with μ_{L} and μ_{S} being the lateral and side friction coefficients, where

$$\mu_{\text{L}} = \mu_{\text{Res}} \frac{S_{\text{L}}}{S_{\text{Res}}}, \quad \mu_{\text{S}} = \mu_{\text{Res}} \frac{S_{\text{S}}}{S_{\text{Res}}}, \tag{12.9}$$

$$\mu_{\text{Res}} = c_1 (1 - e^{c_2 S_{\text{Res}}}) - c_3 S_{\text{Res}}, \quad S_{\text{Res}} = \sqrt{(S_{\text{L}})^2 + (S_{\text{S}})^2}. \tag{12.10}$$

This SimMechanics model offers great versatility, enabling the selection of road types based on their adhesion characteristics (dry asphalt, rain, snow, etc.).

Constraints

The actuator in 4WIMD EV is vital for ensuring control and safety by enforcing constraints on the permissible additional yaw moment. It must operate within these defined limits, regardless of the chosen model, to maintain stability and proper control of the vehicle. In case of the Fox platform, such constraints are

$$\begin{aligned} |\Delta M_z| &\leq 100 \\ |M_z| &\leq 1000, \end{aligned} \tag{12.11}$$

where M_z is the control action, and ΔM_z is its rate of change.

Extensions of Utilizing the Benchmark

The control system aims to optimize vehicle performance in terms of maneuverability and stability. It seeks to achieve responsive and predictable handling characteristics that enhance the driver's control over the vehicle, especially during challenging driving maneuvers, while also ensuring the vehicle maintains tracking for predefined values to enhance stability further.

In this regard, the benchmark for handling double-lane change in vehicle stability control can be adapted to address diverse challenges and scenarios. It might include the variation of dynamic road conditions like curves, slopes, and adverse weather, requiring enhanced control systems to predict and respond effectively. For example, high-speed maneuvers such as emergency lane changes and aggressive cornering may also require optimized stability control algorithms. Likewise, off-road driving scenarios demand adaptations for rough surfaces and obstacles, ensuring stability and maneuverability. Stability control should also be adjusted for emergency scenarios such as rapid braking or loss of traction. These variations can be addressed implementing small modifications in the benchmark so as to tackle a wide range of stability and control challenges across different driving conditions.

12.3 Benchmark Design

The Fox benchmark evaluates stability control in electric vehicles, emphasizing braking technologies to prevent tire locking and slipping, thereby reducing accident risks. As we previously commented, ESC, combined with TVC, enhances stability by addressing oversteering and understeering. These techniques are heavily dependent on the precision and sensitivity of the sensors and actuators used to assess vehicle dynamics and implement control measures. Besides, to incorporate driver variety into the benchmark framework, driver models and tests with several drivers need to be taken into account. Additionally, environmental factors such as road surface conditions and weather fluctuations have significant impacts on vehicle stability and control.

To effectively assess stability control system performance in real-world circumstances, these aspects must be properly accounted for in both experimental setups and simulation models. While other control engineering benchmarks test vehicle stability and control on straight roads or in steady-state cornering, this one stands out as it focuses on double-lane changes, and provides a focused assessment of stability control systems in difficult real-world situations. By simulating double-lane changes, it correctly evaluates the system's capacity to manage quick lateral maneuvers while maintaining stability and control in dynamic environments.

12.3.1 Benchmark Hardware

This benchmark based on the Fox vehicle, illustrated in Fig. 12.1, was built on the SilverCars S2 racing chassis, which got modifications to accommodate the installation of four brushless motors within the wheel wells. Additionally, power sources and monitoring screens were integrated, facilitating effective communication via the CAN bus, a commonly employed protocol within the automotive industry. The car components are detailed as follows:

- **Motors:** The vehicle has four 7 kW brushless hub motors directly connected to the wheels. They are controlled by power converters, whose main specifications are controller supply voltage range (18–90 V), analog brake, and throttle input (0–5 V). An inverter converts battery pack DC to AC for precise speed and torque control. The control unit manages motor responses based on driver input. There is a separate braking system operated by a dedicated pedal for independent control, separate from the motor control.
- **Batteries:** The vehicle's power supply consists of six packs, each containing four cells utilizing lithium-ion technology with LiFeMnPO4 as the cathode material. Their main specifications are nominal voltage (12.8 V), nominal capacity (100 Ah), and operation voltage range (11.2–14.4 V).
- **Sensors:** It has pedal position sensors for precise measurement, a steering wheel rotation sensor for steering angle, and sensors on the accelerator and brake pedals for driver input. Suspension position sensors monitor shock absorber compression, aiding in estimating vertical forces and enhancing vehicle dynamics control. The IMU is a crucial navigation sensor, measuring linear and angular accelerations across three local reference axes. It also includes an integrated GPS sensor for trajectory visualization and potential journey improvements.
- **Control Unit:** It operates with LabVIEW software, managing CAN bus communication and high-level tasks. The Fox vehicle's internal architecture, including the ECU, IMU, electric motors, and communication interfaces. The ECU uses PC104 modules and the QNX real-time operating system, requiring code adaptation for implementation in the vehicle.

12.3.2 Input and Output Data

The data gathered in the benchmark study comprises side slip angle, yaw rate, yaw rate derivative, pedal position, steering wheel rotation, and suspension position. In addition to these sensor readings, data relevant to the performance of the control systems, including torque exerted by the motors and trajectory deviations, are recorded. Inputs to the software include environmental factors (e.g., road conditions, weather), vehicle parameters (e.g., mass, inertia), and driver inputs (e.g., throttle, brake). Outputs include control commands sent to actuators (e.g., motor torque commands, steering angle commands) and feedback signals indicating the vehicle's state (e.g., velocity, acceleration).

12.3.3 Control System and Challenges Associated

The complicated process of calculating tire–road friction poses a significant challenge, as differences in this interaction may significantly affect vehicle behavior. Additionally, the control challenges of the benchmark include particular tests to measure the performance of the stability control system. These tests involve double-lane changes, high-speed cornering, emergency braking, and driving on a variety of road conditions. Evaluating the system's response to changes in driver inputs is also essential. These challenges highlight the importance of careful consideration in experimental settings and simulation models to correctly assess stability control system performance in real-world situations.

12.3.4 Test Cases

The benchmark for evaluating vehicle lateral stability enhancement during double-lane changes includes various specific test cases, each designed to provide in-depth insights. One such test is the standard double-lane change maneuver, where two consecutive lane changes are executed within a predefined distance, assessing the vehicle's responsiveness and stability during rapid lateral transitions. The ISO 3888 standard defines essential tests like the double-lane change dual track maneuver, evaluating a vehicle's lateral dynamics. This involves accelerating the vehicle to a specific speed and navigating through cones without hitting them, thus assessing stability during high-speed curves and emergency situations as in Fig. 12.6.

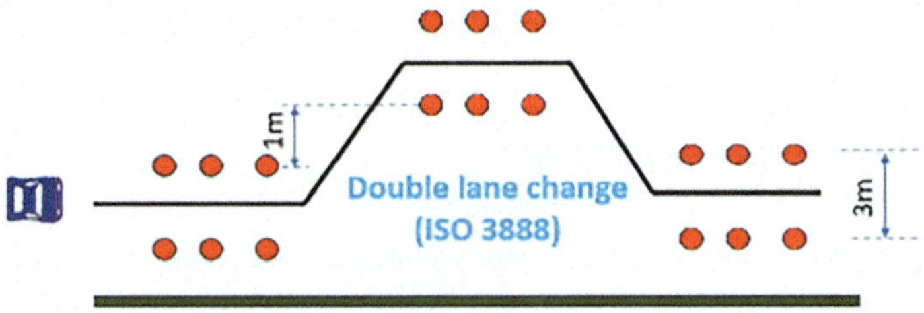

Fig. 12.6 The double-lane change (ISO 3888)

Furthermore, the benchmark includes adaptability testing, which involves adjusting lane widths to imitate various road designs and environments, allowing for a study of the system's performance under varied scenarios. Moreover, stability testing in simulated severe weather conditions, such as rain or snow, evaluates the system's durability and performance in difficult situations. Detailed measurements are gathered during each test scenario, including metrics such as side slip angles, yaw rate, trajectory deviation, and lateral acceleration, to provide complete data for analyzing the vehicle's dynamic behavior and stability control performance.

12.3.5 Performance Metrics

Performance metrics in this benchmark assess responsiveness, stability, and maneuver precision, collectively evaluating the vehicle's agility and accuracy in executing maneuvers. Various metrics are used to comprehensively evaluate control strategies: comparing desired and actual trajectories assesses the vehicle's responsiveness to control inputs and course accuracy, typically quantified by calculating the Root Mean Square Error (RMSE) to quantify deviation. Similarly, comparing current and desired values of angular velocity and sliding angle aids in identifying deviations and refining control strategies, measured using RMSE. The generation of references for both side slip angle and yaw rate is determined as follows. In order to enhance stability and enhance tracking performance, it is recommended to establish the desired side slip angle as zero [14], i.e.,

$$\beta_{\text{des}} = 0. \tag{12.12}$$

Furthermore, supposing that the vehicle is following a curvilinear path with a consistent turning radius R, speed V, a constant angular velocity $\dot{r} = 0$, and a side slip angle rate of change $\dot{\beta} = 0$ [10], the necessary yaw rate can be expressed as

$$r_{\text{des}} = \frac{V}{L + \frac{m}{L}(\frac{l_{\text{r}}}{C_{\alpha\text{f}}} - \frac{l_{\text{f}}}{C_{\alpha r}})V^2}\delta_{\text{f}}. \tag{12.13}$$

This equation establishes the connection between the yaw rate and the front wheel angle of rotation when the vehicle is at a standstill.

Vehicle side acceleration reflects dynamic stability during maneuvers, while monitoring steering wheel angle and analyzing its relationship with angular velocity assess driver input and vehicle agility. Speed fluctuations during maneuvers help assess control system impact on performance, offering insights for improvement. Researchers use these metrics to understand control system effectiveness and refine strategies for safe navigation in various driving conditions.

12.4 Accessing the Benchmark

Accessing the 4WIMD EV benchmark can be provided through direct communication with the authors. Researchers are encouraged to utilize SimMechanics and MATLAB® for simulations. Remote participation is also possible by establishing direct communication channels with the authors. Furthermore, researchers are encouraged to directly engage with the laboratory at the University of Seville for real-world experimentation opportunities with the physical 4WIMD EV system. It is recommended to use MATLAB® from 2018 onwards and to have a computer with optimal specifications for enhanced performance, including 16 GB RAM, a 2.6 GHz Intel Core i7 processor, and a 512 GB SSD.

12.5 Discussion for Future Comparison

DYC and TVC offer precise control over vehicle dynamics but often require sophisticated algorithms and sensor systems to effectively manage lateral dynamics for stability, particularly during sharp maneuvers. Both understeering (loss of grip at the front tires) and oversteering (loss of grip at the rear tires) are significant concerns for EVs. To address these challenges, researchers have developed a range of control techniques aimed at enhancing vehicle handling and stability. These techniques encompass various approaches, each with its unique strengths and weaknesses. For instance, Proportional–Integral–Derivative (PID) controllers are known for their simplicity but may lack robustness, especially in nonlinear systems like EVs. Sliding Mode Control (SMC) offers robustness but may result in chattering and high control effort [1]. Nonlinear Model Predictive Control (NMPC) offers superior performance but demands accurate models and computational intensity [9]. Machine Learning (ML) techniques provide adaptability but can be data-intensive [13]. In the last years, integration strategies combining advanced techniques like MPC and ML algorithms have emerged. This way, MPC optimizes trajectory and performance by incorporating predictive models, while ML dynamically adapts parameters to ensure effective vehicle handling in changing conditions. The integration of ML techniques has notably enhanced MPC's computational efficiency via data-driven modeling and optimization, leading to various integration categories such as offline modeling, online learning, ML replicating MPC behavior, ML within MPC's control structure, and MPC as a safe learning controller.

In connection with the proposed benchmark, in [11] the nonlinear model of the Fox vehicle was presented and validated, [6] focuses on dynamics control in electric vehicles with wheel motors, [4] addresses the control of independent 4WD electric vehicles using the DYC method, and [7, 15] focus on designing traction control strategies to optimize performance by managing power delivery, enhancing traction and stability, reducing wheel slip through sensor data analysis and adaptive algorithms, and [2, 3] evaluate predictive control strategies for maintaining lateral stability in 4WIMD EVs and explore the potential of neural network-based MPC to further enhance lateral stability. Moreover, [8] highlights the potential of advanced torque vectoring systems in modern automotive design.

These studies provide comprehensive insights into the successful deployment of benchmarks in various scenarios, including navigating sharp turns, handling acceleration, and managing deceleration. They not only underscore the benchmark's versatility but also contribute to a deeper understanding of the complex dynamics and efficacy of different control strategies under diverse conditions.

While existing approaches demonstrate strengths in enhancing vehicle responsiveness, stability, and precision during maneuvers, challenges persist in achieving optimal control under diverse real-world conditions. Limitations include potential discrepancies between simulation and real-world performance, the necessity for accurate sensor data, and the complexity of modeling tire–road friction for precise control.

Research Gaps and Opportunities

Despite advancements in enhancing lateral stability in EVs, there are still significant gaps in the current state of the art. One critical gap is the limited availability of standardized benchmarks and evaluation criteria for comparing various control strategies, a limitation we aim to highlight in this chapter. While the discussed benchmark addresses assessing lateral stability during double-lane change, further research is needed to validate its effectiveness across different EV platforms and scenarios.

Additionally, more comprehensive studies are necessary to integrate passive safety systems with active control techniques to enhance vehicle safety and stability. Furthermore, challenges persist in dynamically managing oversteering and understeering during cornering maneuvers, alongside gaps in the adaptability and robustness of stability control systems across diverse road conditions, including extreme weather scenarios.

Integrating safety systems with emerging technologies like LIDAR and computer vision systems holds promise for enhancing EV safety in the medium to long term. However, standardized protocols and benchmarks will be crucial for comprehensively validating their performance. Addressing these existing gaps will ensure continued advancements in EV safety, performance, and user experience, leading to safer and more reliable electric vehicles for drivers.

Acknowledgements The authors would like to thank the projects PID2020-119476RB-I00 and PID2023-152876OB-I00 funded by MCIN/AEI /10.13039/501100011033.

References

1. Guo J, Wang J (2017) Lateral stability control of distributed drive electric vehicle based on fuzzy sliding mode contro. In: 2017 IEEE 3rd information technology and mechatronics engineering conference (ITOEC). IEEE, pp 675–680
2. Hassan A, Frejo JRD, Maestre JM (2023) An assessment of predictive control strategies for lateral stability control of 4-wheels drive electrical vehicle. In: 2023 8th international conference on control and robotics engineering (ICCRE). IEEE, pp 177–182
3. Hassan A, Ruiz-Moreno S, Frejo JRD, Maestre JM, Camacho EF (2024) Neural-network based MPC for enhanced lateral stability in electric vehicles. IEEE Access
4. Hernández G, Bordons Alba C, Marcos D, Montero C (2015). Control de Estabilidad basado en MPC para Un vehículo eléctrico Con Motores en Rueda. Jornadas de Automática 887–894
5. Huang B, Sen W, Huang S, Xiang F (2018) Lateral stability control of four-wheel independent drive electric vehicles based on model predictive control. Math Probl Eng 1–15:2018
6. Kanchwala H, Wideberg J, Alba CB, Marcos D (2015) Control of an independent 4WD electric vehicle by DYC method. Int J Vehicle Syst Model Test 10(2):168–184
7. Montero C, Marcos D, Bordons C, Ridao MA, Camacho EF, Gonzalez E, Oliva A (2015) Modeling and torque control for a 4-wheel-drive electric vehicle. In: 2015 IEEE international conference on industrial technology (ICIT). IEEE, pp 2650–2655
8. Morera E, Ocampo-Martinez C, Bianchi F (2022) Experimental modelling and optimal torque vectoring control for 4WD vehicles. IEEE Trans Veh Technol 71(5):4922–4932
9. Parra A, Tavernini D, Gruber P, Sorniotti A, Zubizarreta A, Pérez J (2020) On nonlinear model predictive control for energy-efficient torque-vectoring. IEEE Trans Veh Technol 70(1):173–188
10. Rajamani R (2011) Vehicle dynamics and control. Springer Science & Business Media
11. Rodríguez DM (2014) Contributions to power management and dynamics control in hybrid vehicles. PhD thesis, Universidad de Sevilla
12. Sanders KT, Webber ME (2012) Evaluating the energy consumed for water use in the United States. Environ Res Lett 7(3):034034
13. Vasiljevic G, Bogdan S (2016) Model predictive control based torque vectoring algorithm for electric car with independent drives. In: 2016 24th Mediterranean conference on control and automation (MED). IEEE, pp 316–321
14. Wang H, Huang Y, Khajepour A, Zhang Y, Rasekhipour Y, Cao D (2019) Crash mitigation in motion planning for autonomous vehicles. IEEE Trans Intell Transp Syst 20(9):3313–3323
15. Wideberg J, Bordons C, Luque P, Mántaras DA, Marcos D, Kanchwala H(2014) Development and experimental validation of a dynamic model for electric vehicle with in hub motors. Procedia-Social Behav Sci 160:84–91

Chapter 13
Benchmark Problem for Visual Nudge: Light-Guided Control of Human-Driven Vehicles

Masaki Inoue, Masahiko Takeda, and José M. Maestre

13.1 Introduction

Addressing traffic congestion is a multifaceted challenge, encompassing the mitigation of environmental impacts, reduction of economic losses, and the lowering of the risk of traffic accidents. Particularly in highways, the traffic congestion is mainly caused by the unconscious driving behavior of human drivers in excessive traffic demand as pointed out in [5]. For instance, at the entrance of a tunnel, drivers unconsciously release the accelerator due to changes in brightness. The deceleration of a single vehicle, cascading through following vehicles, leads to traffic congestion. Various trials have been made to mitigate traffic congestion. Traffic congestion could be alleviated in the future when autonomous vehicles have a significant market share, as they can be appropriately controlled automatically. However, in the near term, the majority of vehicles are manually driven by human drivers, necessitating congestion relief strategies that focus on such human-driven vehicles (HDVs).

The approach to HDVs addressed in this chapter is to use pace-making-lights (PMLs), which are multiple lights installed inside highway tunnels, as illustrated in Fig. 13.1. Drivers tend to unconsciously follow objects moving faster than themselves. PMLs leverage this driver tendency by sequentially illuminating multiple

Masaki Inoue and Masahiko Takeda (equally contributed to this work).

M. Inoue · M. Takeda
Keio University, 3-14-1 Hiyoshi, Kohoku-ku, Yokohama, Kanagawa, Japan
e-mail: minoue@appi.keio.ac.jp

M. Takeda
e-mail: masahiko0322@keio.jp

J. M. Maestre (✉)
University of Seville, Camino de los Descubrimientos sn., 41092 Sevilla, Spain
e-mail: pepemaestre@us.es

J. M. Maestre and C. Ocampo-Martinez (eds.), *Control Systems Benchmarks*, Advances in Industrial Control, https://doi.org/10.1007/978-3-031-76312-0_13

Fig. 13.1 Pace-making-lights installed in a Tomei Highway tunnel, Japan

lights in the direction of vehicle movement at a speed faster than the vehicle, thereby promoting the suppression of vehicle speed reduction [10]. The design problem of PML dynamics, which operates depending on the vehicle's state, can be viewed as a feedback controller design problem. Particularly, it represents a control problem in a human-in-the-loop control system, extending beyond automatic control to include the driver in the feedback loop.

This chapter presents a benchmark problem for the human-in-the-loop control design, in particular, the design of the PML dynamics for HDVs, which explores control stimuli for human drivers to reach the desired vehicle speed. This benchmark problem includes the driver model. When integrated with the vehicle dynamics model, the driver model simulates how acceleration and deceleration respond to the behavior of the PML and preceding vehicles. The incorporation of the HDV model in the control benchmark enables the evaluation of the performance of advanced control methods such as model predictive control (MPC, [6]), robust control [11], data-driven control [3], and their recent advancements.

13.2 Problem Description

13.2.1 Dynamics of Human-Driven Vehicle

Recall that the plant system to be controlled in this chapter is the HDV. As illustrated in Fig. 13.2, the control target HDV is denoted by P, and the human driver accelerates or decelerates P depending on the state of the PMLs and the preceding vehicle P_{pv}.

The dynamics of the HDV P is composed of the vehicle's mechanical dynamics V and human driver's dynamics H, expressed as $P = (V, H)$ and illustrated as Fig. 13.3. The vehicle dynamics V is modeled by the first-order system with the

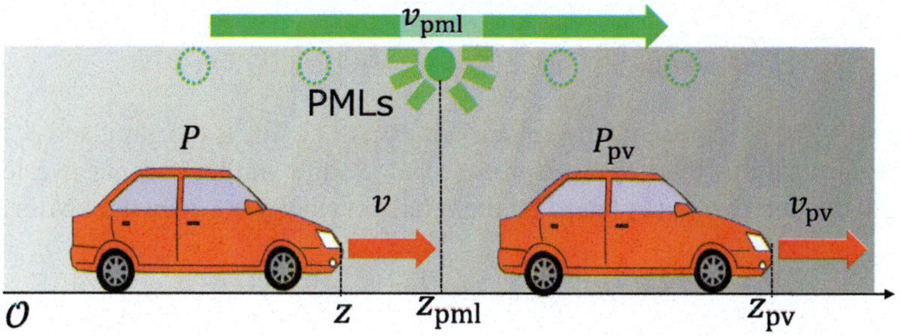

Fig. 13.2 Human-driven vehicle to be controlled P, preceding vehicle $P_{\rm pv}$, and PMLs

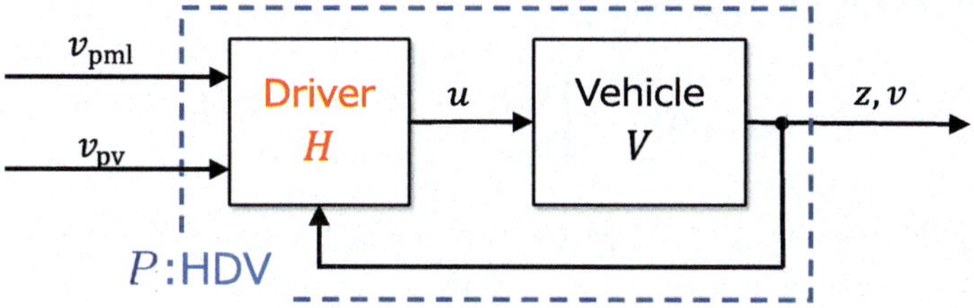

Fig. 13.3 Internal structure of human-driven vehicle P

double integrator and is described as

$$V: \begin{cases} z(k+1) = z(k) + \Delta t v(k), \\ v(k+1) = v(k) + \Delta t a(k), \\ a(k+1) = \{(T - \Delta t)/T\}a(k) + (\Delta t/T)\mathrm{sat}_{u_{\max}}(u(k)), \end{cases} \tag{13.1}$$

where $z(k), v(k), a(k), u(k) \in \mathbb{R}$ represent the position, velocity, acceleration, and the driver's accelerator input, respectively, $\Delta t \in \mathbb{R}_+$ is the sampling time, and $T \in \mathbb{R}_+$ is the time constant. In addition, symbol $\mathrm{sat}_*(\cdot)$ represents the saturation function. Letting $u_{\max} \in \mathbb{R}_+$ be a constant that represents the bound of the acceleration input, and $\mathrm{sat}_{u_{\max}}(u)$ is defined by

$$\mathrm{sat}_{u_{\max}}(u) = \begin{cases} u_{\max}, & u \in [u_{\max}, \infty), \\ u, & u \in [-u_{\max}, u_{\max}), \\ -u_{\max}, & u \in (-\infty, -u_{\max}). \end{cases}$$

The driver determines the accelerator input $u(k)$ based on the distance and relative velocity between the preceding vehicle $P_{\rm pv}$, the PML, and his/her own HDV P. Furthermore, it is assumed that the driver model H is modeled by a switched system composed of the preceding vehicle-dependent and the PML-dependent parts, denoted by $H_{\rm pv}$ and $H_{\rm pml}$, respectively. Let $H_{\rm pv}$ generate the output $u_{\rm pv}(k)$ based on the relative velocity between $P_{\rm pv}$ and P when they are sufficiently apart, indicating a low risk of collision:

$$H_{\rm pv} : u_{\rm pv}(k) = c_{\rm pv}\left(v_{\rm pv}(k - d_{\rm pv}) - v(k - d_{\rm pv})\right), \tag{13.2}$$

where $v_{\rm pv}(k) \in \mathbb{R}$ represents the velocity of $P_{\rm pv}$, $c_{\rm pv} \in \mathbb{R}_+$ is the gain, and $d_{\rm pv} \in \mathbb{R}_+$ is the time delay of the driver response to $P_{\rm pv}$.[1] In a similar manner, we let $H_{\rm pml}$ generate the output $u_{\rm pml}(k)$ based on the relative velocity between the PMLs and P:

$$H_{\rm pml} : u_{\rm pml}(k) = c_{\rm pml}\left(v_{\rm pml}(k - d_{\rm pml}) - v(k - d_{\rm pml})\right),$$

where $v_{\rm pml}(k) \in \mathbb{R}$ represents the velocity of the PMLs, $c_{\rm pml} \in \mathbb{R}_+$ is the gain, and $d_{\rm pml} \in \mathbb{R}_+$ is the time-delay of the driver response to the PMLs. Then, the dynamics of the human driver H is modeled as

$$H : u(k) = \begin{cases} u_{\rm pml}(k), & s_{\rm pml}(k) = 1, \\ u_{\rm pv}(k), & s_{\rm pml}(k) = 0, \end{cases} \tag{13.3}$$

where $s_{\rm pml}(k) \in \{0, 1\}$ is a binary variable defined as follows: $s_{\rm pml}(k) = 1$ if the PMLs are activated at time k, $s_{\rm pml}(k) = 0$ if they are not.

The overall dynamics of the HDV $P = (V, H)$ is modeled as follows. Let x be a vector composed of z, v, and a, i.e.,

$$x(k) = \left[z(k)\ v(k)\ a(k)\right]^{\top}.$$

Then, the combination of V in (13.1) and H in (13.3) is reduced to the following state equation:

$$P : \begin{cases} x(k+1) = Ax(k) + B{\rm sat}_{u_{\max}}(u(k)), \\ u(k) = \begin{cases} u_{\rm pml}(k), & s_{\rm pml}(k) = 1, \\ u_{\rm pv}(k), & s_{\rm pml}(k) = 0, \end{cases} \end{cases} \tag{13.4}$$

where A and B are system matrices given by

$$A = \begin{bmatrix} 1 & \Delta t & 0 \\ 0 & 1 & \Delta t \\ 0 & 0 & \frac{T - \Delta t}{T} \end{bmatrix}, \quad B = \begin{bmatrix} 0 \\ 0 \\ \frac{\Delta t}{T} \end{bmatrix},$$

respectively.

[1] Note that the model of $H_{\rm pv}$ is a highly simplified representation and is not valid when the distance between the HDV P and the preceding vehicle $P_{\rm pv}$ is either very far or very close. For instance, in close proximity, the driver's behavior depends not only on relative velocity, denoted by $v_{\rm pv} - v$, but also on relative distance, denoted by $z_{\rm pv} - z$. Conversely, when the vehicles are significantly separated, the driver takes driving actions without particular concern for the state of $P_{\rm pv}$. Therefore, the model given in (13.2) is valid if $z_{\rm pv} - z \in [L_1, L_2]$ holds for some suitable constants L_1 and L_2.

The parameters in the HDV model (13.4) are given as follows. The parameters in the vehicle dynamics V are fixed at $\Delta t = 0.20$ s, $T = 0.25$ s, and $u_{\text{max}} = 4.00$. Notice that the model parameters in the driver's dynamics are expected to vary individually. Reasonable initial parameter values to start the tuning procedure are $c_{\text{pv}} = 0.258$, $d_{\text{pv}} = 7$, $c_{\text{pml}} = 0.719$ and $d_{\text{pml}} = 2$. These values were determined by the second author based on his experimental data during the setup process of the VR experiment environment. For simplicity of notation, we let $P_i, i \in \{1, 2, \ldots, N\}$ denote the HDV model (13.4) with the parameters of driver No. i.

13.2.2 Problem Description of PML-Based Control

In the benchmark, we address feedback control problems that involve designing the PML-based control logic, denoted by K. It is assumed that the velocities of the HDV P and the preceding vehicle P_{pv}, denoted by $v(k)$ and $v_{\text{pv}}(k)$, respectively, are measurable in real time. Then, we manipulate the velocity of the PMLs, denoted by $v_{\text{pml}}(k)$, to ensure that the HDV quickly reaches its desired velocity, typically matching that of the preceding vehicle as $v(k) \to v_{\text{pv}}(k)$. To this end, we describe the dynamics of the PMLs to be designed as

$$K : v_{\text{pml}}(k) = \mathcal{K}(v(k), v_{\text{pv}}(k)), \tag{13.5}$$

where $\mathcal{K}$ represents a dynamic operator, i.e., $v_{\text{pml}}(k)$ can depend on its past state such as $v_{\text{pml}}(k-1)$. Then, we address the design of the overall control system, denoted by (P, K), with the objective of achieving $v(k) \to v_{\text{pv}}(k)$.

The detailed setup of the benchmark is as follows. The initial state of P is set as

$$z(0) = 0, \quad v(0) = 70.$$

The state behavior of P_{pv} is generated from

$$z_{\text{pv}}(0) = 10, \quad v_{\text{pv}}(k) = \begin{cases} 70, & k \in [0, k_{\text{start}}), \\ 80, & k \in [k_{\text{start}}, k_{\text{end}}], \end{cases}$$

where $k_{\text{start}} = 5/\Delta t = 25$ and $k_{\text{end}} = k_{\text{start}} + 15/\Delta t = 100$ represent the discrete times when the PMLs are turned on and when the benchmark simulation is terminated, respectively. The control scenario, featuring the preceding vehicle's behavior $v_{\text{pv}}(k)$ along with a typical example of the PMLs behavior $v_{\text{pml}}(k)$ and the resulting HDV behavior $v(k)$, is given in Fig. 13.4. We see in the figure that the preceding vehicle suddenly accelerates, and the PMLs prompt the HDV to track the preceding vehicle. This situation is analogous to simulating the following scenario: supposing that the HDV decelerates and reduces its velocity at the entrance of a tunnel, we employ suitable PML-based control to restore the HDV's velocity.

In the benchmark, we address the following two control problems.

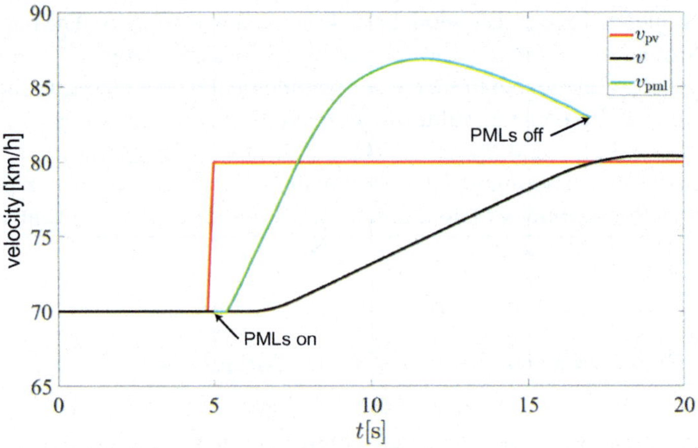

Fig. 13.4 Typical control scenario of featuring the behavior of the preceding vehicle, denoted by v_{pv}, the PMLs behavior v_{pml}, and resulting HDV behavior v

Problem 13.1 *Tracking control problem for specific HDV*: we design the PML-based control logic K, given in (13.5), to ensure the velocity $v(k)$ of the control system (P_i, K), defined for a *specific* driver $i \in \{1, 2, \ldots, N\}$, which quickly tracks the velocity $v_{pv}(k)$ of the preceding vehicle P_{pv}.

Problem 13.2 *Robust tracking control problem for all HDVs*: we design the PML-based control logic K of (13.5) to ensure $v(k)$ of every control system (P_i, K), $i \in \{1, 2, \ldots, N\}$ which quickly tracks $v_{pv}(k)$.

In addition to general description stated as in Problems 13.1 and 13.2, several physical and safety specifications must be satisfied. One of the specifications pertains to the constraints on the driver's accelerator input $u(k)$, as is included in the HDV model in (13.4). Physically, the accelerator input must satisfy $u(k) \in [-u_{max}, u_{max}]$. Furthermore, it is essential to maintain a suitable following distance from the preceding vehicle. This is modeled by an inequality constraint such as $z_{pv}(k) - z(k) \geq \mu$ for some positive constant μ. It should be emphasized again that the PML-based control must embody the concept of a *visual nudge*. This implies weakly guiding human drivers rather than enforcing control. PMLs provide visual stimuli to the drivers, indirectly influencing the vehicle through their reactions. Consequently, we need to prioritize the comfort of human drivers and minimize any potential discomfort. To this end, the control problem should encompass system specifications aimed at reducing frequent changes in PMLs' velocity. This is modeled as $|v_{pml}(k+1) - v_{pml}(k)| \leq \rho$ for some positive constant ρ.

A fundamental approach to address Problems 13.1 and 13.2 is by employing model-based control. This approach implicitly assumes the availability of individual HDV models P_i and/or their averaged representation, enabling the utilization of advanced control techniques such as MPC and robust control. It is also possible to explore an alternative approach involving data-driven control. For instance, this alternative approach considers scenarios where the parameters of the driver model (13.3) are unknown. Instead, data such as the HDV state behavior $\{z(k), v(k)\}$, the PMLs' velocity $\{v_{\rm pml}(k)\}$, as well as data on the preceding vehicle's velocity $\{v_{\rm pv}(k)\}$, are measurable in real time. In such instances, the control logic K estimates model parameters using this data and it updates the PMLs' velocity $v_{\rm pml}(k+1)$. This approach provides the flexibility to address Problems 13.1 and 13.2 effectively.

13.3 Benchmark Design

Recall that the problem description for the benchmark is stated in Problems 13.1 and 13.2 of Sect. 13.2.2. In the benchmark simulation, data on the HDV state behavior $\{z(k), v(k)\}$ and data on the PMLs' velocity $\{v_{\rm pml}(k)\}$ are collected. Based on the data, the control performance is evaluated.

The metric of evaluating the tracking performance of designed control logic K is given as follows: letting $v_i(k)$ denote the HDV's velocity generated by (P_i, K), which depends on the model of driver i, the metric for Problem 13.1 is given by

$$J_{(13.1)} = \frac{1}{k_{\rm end} - k_{\rm start} + 1} \left(\sum_{k=k_{\rm start}}^{k_{\rm end}} (v_{\rm pv}(k) - v_i(k))^2 \right)^{\frac{1}{2}},$$

which is applied for each HDV model with driver i.

In an extensive manner, the metric for Problem 13.2, of evaluating the *worst-case* tracking performance, is described as

$$J_{(13.2),{\rm ro}} = \frac{1}{k_{\rm end} - k_{\rm start} + 1} \max_{i \in \{1,2,\ldots,N\}} \left(\sum_{k=k_{\rm start}}^{k_{\rm end}} (v_{\rm pv}(k) - v_i(k))^2 \right)^{\frac{1}{2}},$$

which is based on the simulation with the HDV model (13.4) with all drivers $i \in \{1, 2, \ldots, N\}$.

An alternative metric for Problem 13.2 is of evaluating the *average* tracking performance and is described by

$$J_{(13.2),{\rm av}} = \frac{1}{k_{\rm end} - k_{\rm start} + 1} \frac{1}{N} \sum_{i=1}^{N} \left(\sum_{k=k_{\rm start}}^{k_{\rm end}} (v_{\rm pv}(k) - v_i(k))^2 \right)^{\frac{1}{2}}.$$

Since it relies on human drivers' unconscious reactions, there is inevitably some uncertainty in each HDV behavior, expressed as $v_i(k)$. To mitigate this uncertainty and enhance the performance in terms of metrics $J_{(13.2),{\rm ro}}$ or $J_{(13.2),{\rm av}}$, the control logic K should exhibit robustness against model uncertainties.

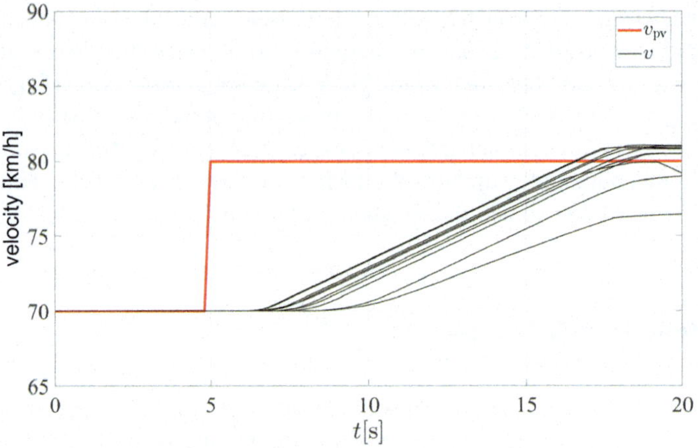

Fig. 13.5 Demonstration of simulating ten HDVs

The specific control scenario illustrated in Fig. 13.4 is designed based on MPC (see the work by the authors in [9] for details) and its approximation to a machine learning model. The MPC control logic is designed based on the averaged model of all HDV models (P_i, K), $i \in \{1, 2, \ldots, N\}$. Figure 13.5 shows the result of the test case, utilizing an MPC-based control logic in K. In the figure, the red line represents $v_{\mathrm{pv}}(k)$, i.e., the velocity of the preceding vehicle P_{pv}, while the black ones represent $v_i(k)$, $i \in \{1, 2, \ldots, N\}$, i.e., the velocity of the controlled HDV (P_i, K) for all $i \in \{1, 2, \ldots, N\}$. We observe that each $v_i(k)$ approaches $v_{\mathrm{pv}}(k)$ even when their individual trajectories differ.

This benchmark has been constructed to simulate a real system. However, implementing this in a real system, while theoretically possible, is not a straightforward task. It would require the installation of PMLs in the traffic infrastructure and the establishment of an environment where they can be controlled in real time. Furthermore, the assumption made in the benchmark, that the position and velocity of the HDVs are measurable in real time, is practically challenging. If these challenges are overcome, it is expected that the PML-based control technology can contribute to the development of advanced transportation systems.

13.4 Access to Benchmark

The benchmark for the PML design is accessible on GitHub,[2] providing MATLAB® and Unity codes that include specific parameters of the HDV model (13.4). The main part of the benchmark is in the MATLAB® codes, facilitating control demonstrations for HDV models under the PML-based control, denoted as (P_i, K). The resulting control performance is assessed for each demonstration using J_1, $J_{(13.2),\mathrm{ro}}$, and $J_{13.2,\mathrm{av}}$.

Furthermore, the control demonstrations, based on the Unity codes, offer the opportunity to experience them in a VR environment. Users can optionally employ an Oculus Quest 2 to immerse themselves in the HDV's behavior from a driver's view in the VR setting (see Fig. 13.6). Examples of both a driver's view and an external view are provided in Figs. 13.7 and 13.8, respectively. The flowchart illustrating the usage of the benchmark is presented in Fig. 13.9. In benchmark.m, users have the flexibility to update and replace the control logic by their own one. Then, they need to select the mode from "all" or "each," which signifies either robust control simulation for all ten HDV models or nominal control simulation for a specific HDV model chosen from No.1 to No.10.

Fig. 13.6 Immersion in VR environment

[2] https://github.com/Xinrui-Fang/Virtual-Car-Project.
Additionally, the detailed documentation of the benchmark and a video demonstrating the driver's view in VR experiments are also accessible on GitHub to assist users with the benchmark's setup and usage.

Fig. 13.7 Driver view in VR environment

Fig. 13.8 External view in VR environment

13.5 Discussion for Future Comparison

13.5.1 MPC-Based Approach to Benchmark

In this subsection, an example of the PML-based control logic is presented. In the work by the authors [9], the control logic K employs MPC and its approximation by a machine learning model. The brief description of the MPC-based control logic is given as follows.

Recall here the control specifications for the benchmark, including the minimization of driver discomfort. The specification is modeled as reducing the ℓ_1 norm of the time sequences of $\{v_{\mathrm{pml}}(k+1) - v_{\mathrm{pml}}(k)\}$. See the work [7] for the ℓ_1 optimal control, which promotes the sparsity. This norm constraint is utilized for the following MPC formulation.

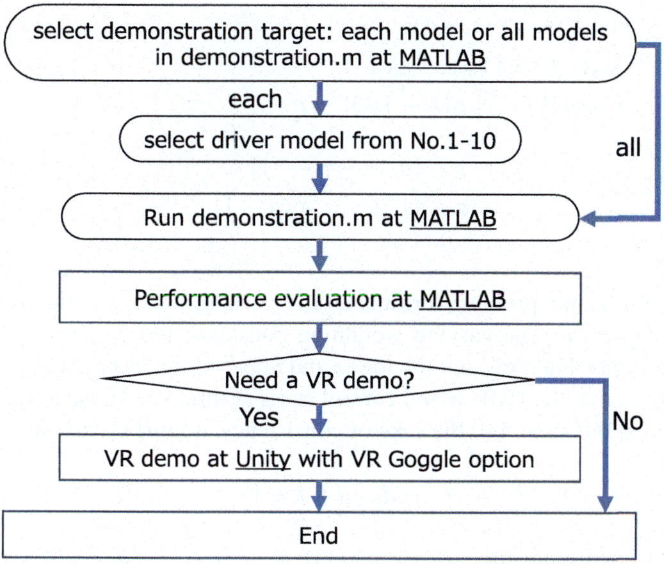

Fig. 13.9 Flowchart of benchmark

Consider that the MPC prediction horizon is denoted as τ_H and the control horizon is one. We assume that $x(k)$, $v(k)$, and $v_{\rm pv}(k)$ are measurable and that $v_{\rm pml}(k-d_{\rm pml})$, $v_{\rm pml}(k-d_{\rm pml}+1), \cdots, v_{\rm pml}(k-1)$ and $v(k-d_{\rm pml}), v(k-d_{\rm pml}+1), \cdots, v(k-1)$ are stored in K at time k. Furthermore, we consider that the preceding vehicle does not accelerate nor decelerate, i.e., it holds that

$$v_{\rm pv}(k+\tau+1) = v_{\rm pv}(k),\ \tau \in \{0, \cdots, \tau_H - 1\}.$$

To clarify the following MPC formulation, for any time sequence $z(k)$, we denote its future prediction at τ steps ahead from the current time k as $z(\tau|k)$. Furthermore, for simplicity of notation, we let $\{v_{\rm pml}(\tau|k)\} = \{v_{\rm pml}(0|k), v_{\rm pml}(1|k), \ldots, v_{\rm pml}(\tau_H|k)\}$. Then, recalling the state variable of the HDV as $x(k) = [\, z(k)\ v(k)\ a(k)\,]$, the optimization problem solved at each time step k is formulated as follows:

$$\min_{\{v_{\rm pml}(\tau|k)\}} \sum_{\tau=0}^{\tau_H-1} \left(q_\tau (v(\tau+1|k) - v_{\rm pv}(k))^2 + r_\tau |v_{\rm pml}(\tau|k) - v_{\rm pml}(\tau-1|k)|\right) \tag{13.6a}$$

$$\text{sub. to}\quad x(\tau+1|k) = Ax(\tau|k) + B\,\mathrm{sat}_{u_{\max}}(u(\tau|k)), \tag{13.6b}$$

$$x(0|k) = x(k), \tag{13.6c}$$

$$u(\tau|k) = c_{\rm pml}\left(v_{\rm pml}(\tau-d_{\rm pml}|k) - v(\tau-d_{\rm pml}|k)\right), \tag{13.6d}$$

$$v_{\rm pml}(-d|k) = v_{\rm pml}(k-d), \tag{13.6e}$$

$$v(-d|k) = v(k-d), \tag{13.6f}$$

$$v_{\text{pml}}(\tau|k) \in [\, v_{\min},\ v_{\max}\,], \tag{13.6g}$$

$$|v_{\text{pml}}(\tau|k) - v_{\text{pml}}(\tau-1|k)| \le \rho, \tag{13.6h}$$

$$\tau \in \{0, \cdots, \tau_H - 1\},$$
$$d \in \{1, 2, \ldots, d_{\text{pml}} - 1\}.$$

In the optimization problem formulated in (13.6), q_τ and r_τ are the positive constants that represent time-varying weighting constants and $v_{\min}$ and $v_{\max}$ are the positive constants that represent the lower and upper limits of v_{pml}, respectively.

We suppose that the PML-based control starts at time $k = 0$, meaning the PMLs are activated at this time. For the sake of consistency, it must satisfy the condition:

$$v_{\text{pml}}(\tau - d_{\text{pml}}|k) = v(\tau - d_{\text{pml}}|k),\ \tau \in \{0, 1, \ldots, d_{\text{pml}} - 1 - k\},$$

which consequently results in $u(\tau|k) = 0$, $\tau \in \{0, 1, \ldots, d_{\text{pml}} - 1 - k\}$ due to (13.6d). This, in turn, implies that the control logic initiates at $k = 0$ as $v_{\text{pml}}(0|0) \neq 0$, but it begins to influence the driver's actions from $k \geq d_{\text{pml}}$ since $u(0|d_{\text{pml}}) \neq 0$. It is crucial to account for such delays in the driver's reactions when designing the control logic K.

13.5.2 Discussion for Alternative Approaches

Note that the PML-based control, as represented in (13.6), heavily relies on human drivers' unconscious reactions, inevitably leading to some degree of uncertainty in their responses. To address and mitigate this uncertainty, and to enhance performance measured by metrics like J_1, $J_{2,\text{ro}}$, and $J_{2,\text{av}}$, several effective approaches can be considered. The approaches include robust control [11], which explicitly accounts for uncertainty, adaptive control [2], which involves updating the control logic based on estimated driver tendencies, such as the model parameters c_{pml} and d_{pml}, and gain scheduling [8], coupled with parameter estimation mechanisms. Each of the strategies plays a crucial role in enhancing the performance and robustness of the control system, even in the presence of the uncertainty arising from human driver reactions.

Moreover, it is essential to consider compensation methods as supplements to the PML-based control to further ensure the stability and safety of the overall control system. These compensation methods include weak control [4] and control barrier functions [1].

In addition to the extension of the PML-based control through the integration of compensation methods, an alternative approach could be exploring data-driven control, as opposed to model-based methods like (13.6). As mentioned at the end of

Sect. 13.2, the data-driven approach is tailored for scenarios where the driver model (13.3) is not available, which is a reasonable assumption for practical systems. In this approach, the PML-based control logic is updated using operational data on the PMLs' and human drivers' reactions.

Acknowledgements This work was supported by Grant-in-Aid for Scientific Research (B), No. 20H02173 and No. 20H04473 from JSPS. Financial support from the Spanish MCIN/AEI/ 10.13039/501100011033 Project C3PO-R2D2 (Grant PID2020-119476RB-I00). The authors acknowledge Xinrui-Fang for his support in setting up the VR experiments.

References

1. Ames AD, Xu X, Grizzle JW, Tabuada P (2016) Control barrier function based quadratic programs for safety critical systems. IEEE Trans Autom Control 62(8):3861–3876
2. Åström KJ, Wittenmark B (2013) Adaptive control. Courier Corporation
3. Hou ZS, Wang Z (2013) From model-based control to data-driven control: survey, classification and perspective. Inf Sci 235:3–35
4. Inoue M, Gupta V (2019) Weak control for human-in-the-loop systems. IEEE Control Syst Lett 3(2):440–445
5. Koshi M, Kuwahara M, Akahane H (1992) Capacity of sags and tunnels on Japanese motorways. ITE J 62(5):17–22
6. Mayne DQ (2014) Model predictive control: Recent developments and future promise. Automatica 50(12):2967–2986
7. Nagahara M (2020) Sparsity methods for systems and control. Now Publishers
8. Rugh WJ, Shamma JS (2000) Research on gain scheduling. Automatica 36(10):1401–1425
9. Takeda M, Inoue M, Fang X, Minami Y, Maestre J (2022) Light guidance control of human drivers: driver modeling, control system design, and VR experiment. IFAC-PapersOnLine 55(41):32–37
10. Terada H, Yanagihara M, Oneyama H (2021) Influence of moving light guide system on traffic flow in presence of autonomous vehicles. Int J Intell Transp Syst Res 19:335–346
11. Zhou K, Doyle JC, Glover K (1996) Robust and optimal control. Prentice Hall, Upper Saddle River

Index

J. M. Maestre and C. Ocampo-Martinez (eds.), *Control Systems Benchmarks*, Advances in Industrial Control, https://doi.org/10.1007/978-3-031-76312-0

The manufacturer's authorised representative in the EU is Springer Nature Customer Service Centre GmbH, Europaplatz 3, 69115 Heidelberg, Germany. If you have any concerns regarding our products, please contact ProductSafety@springernature.com

Printed and bound by CPI Group (UK) Ltd, Croydon, CR0 4YY
30/09/2026
02225593-0002